Methods of Developing Sliding Mode Controllers

Methods of Developing Sliding Mode Controllers

Design and MATLAB Simulation

Reihaneh Kardehi Moghaddam

Department of Electrical Engineering, Mashhad Branch,
Islamic Azad University, Mashhad, Iran and
School of Computing, Macquarie University, Sydney, Australia

Mostafa Rabbani
Department of Electrical Engineering, Mashhad Branch,
Islamic Azad University, Mashhad, Iran

IEEE Press Series on Control Systems Theory and Applications
Maria Domenica Di Benedetto, Series Editor

IEEE Press

WILEY

Published by John Wiley & Sons, Inc., Hoboken, New Jersey.
Published simultaneously in Canada.

Library of Congress Cataloging-in-Publication Data Applied for:
Hardback ISBN: 9781394314485

Cover Design: Wiley
Cover Images: © sankai/Getty Images, © Pobytov/Getty Images

Set in 9.5/12.5pt STIXTwoText by Straive, Pondicherry, India

To my beloved husband, Mohammad, whose unwavering love and steadfast support have been a constant source of strength and inspiration throughout my journey.

To my parents, Masi and Hassan, for their unwavering love and encouragement throughout my life.

And to my shining stars, Sepehr and Elena, who have illuminated my world with boundless joy, love, and a sense of purpose. May the light of your laughter and the warmth of your hugs continue to inspire my every endeavor.

With eternal love, I dedicate this book to you all.

—Reihaneh Kardehi Moghaddam

To my beloved wife, Fatima, who has stood by my side with unwavering patience and affection.

To my parents, Zahra and Ali, for their unwavering support throughout all phases of my education.

To my children, Mohammad Mahdi and Mahda, whose presence brings me joy and inspires my endeavors.

—Mostafa Rabbani

Contents

About the Authors

Reihaneh Kardehi Moghaddam

Dr. Reihaneh Kardehi Moghaddam is a distinguished academic and researcher, focusing on control of nonlinear uncertain systems, intelligent optimization methods, and intelligent learning methods. She currently holds the position of Honorary Associate Professor at the School of Computing, Macquarie University, Australia, and serves as an associate professor at the Department of Electrical Engineering of the Islamic Azad University, Mashhad, Iran. Her extensive research has led to significant contributions in the aforementioned fields, with industrial applications primarily focused on robotics, renewable energy systems, and biomedical image processing. With 14 years of experience teaching nonlinear controllers and an extensive publication history, Dr. Kardehi Moghaddam's dedication to her research areas has motivated her to author this book.

Mostafa Rabbani

He is a PhD student in control engineering at the Islamic Azad University, Mashhad, Iran. His main research interests include nonlinear control, optimal control, and ocean engineering. He is currently a lecturer at the Islamic Azad University and works as a senior engineer of exploitation operations at Shahid Hasheminejad Gas Processing Company (SGPC).

Acknowledgments

I am deeply grateful to Professor Naser Pariz for his guidance and unwavering support in both my academic journey and personal life. His wisdom and encouragement have significantly impacted my growth, and I will always cherish his valuable advice.

I would like to extend my heartfelt thanks to Professor Amin Beheshti, the head of the Data Science Lab and founder of the Big Data Society at Macquarie University, for his invaluable support and for providing insightful scientific comments that have truly motivated me in my academic pursuits.

I am deeply thankful to the Wiley publishing team for their exceptional support, professionalism, and dedication throughout this project. Their well-organized approach ensured a smooth collaboration and significantly contributed to the book's success. In addition, I would like to express my gratitude to the IEEE Press Editorial Board for their valuable guidance and support throughout the development of this book. Their contributions and expertise have been instrumental in shaping the content and ensuring its quality.

Introduction

This book is the culmination of Reihaneh Kardehi Moghaddam's 14 years of lecturing and researching experience in the design and implementation of nonlinear controllers. With a focus on innovation and practical application, the text provides an accessible and comprehensive guide to understanding the complex concepts of controlling uncertain nonlinear systems and designing sliding mode controllers, which are effectively used in electrical, mechanical, civil, and biomedical engineering.

Unique features set this book apart, making it an invaluable resource for graduate and postgraduate engineering students. Key highlights include:

- Clear and user-friendly MATLAB programming examples
- Simple theoretical explanations to aid comprehension
- Professional figures and tables to enhance understanding
- Coverage of the most recent advancements in sliding mode controllers

By combining theoretical foundations with practical applications, this book serves as an engaging reference for students and researchers alike. It encourages the exploration of cutting-edge techniques and fosters a deeper appreciation for the evolving field of nonlinear controllers.

1

An Introduction to Sliding Mode Controllers

1.1 Introduction

In the world of control systems, sliding mode controller (SMC) is a nonlinear control strategy that uses a control signal with a variable structure to change the dynamics of a nonlinear system that forces the system to "slide" along a predetermined path. It is called the sliding surface, and the curved movement of the system path on this surface will guarantee the stability of the system in the presence of parametric and structural uncertainties with a limited bound. The state feedback control law in sliding mode controllers is not a continuous function in terms of time, and, in other words, the position of the system path curve in the state space is dependent. Hence, sliding mode control is a variable structure control method. This variable structure controller is designed in such a way that the system trajectories always converge toward the sliding surface, and by sliding on this path, the tracking error of the desired behavior in the control system tends to zero. The movement of the system during sliding along the neighborhood of the sliding surface is called sliding mode. In the framework of modern control theory, the SMC variable structure system is considered as a subset of hybrid dynamic control systems, because the path curve of the system moves continuously through different and discrete control laws.

Sliding mode control has two main advantages. The first advantage is that by choosing the appropriate sliding function, the desired dynamic behavior of the system can be achieved. And secondly, the closed-loop response of the system has no sensitivity to uncertainties (model parameters, disturbances, and nonlinearity). As a result, from a practical point of view, nonlinear processes are controlled by using sliding mode control in the presence of model disturbances and uncertainties. Unique features such as accuracy, consistency, and easy setup and implementation have made sliding controllers very popular.

Methods of Developing Sliding Mode Controllers: Design and MATLAB Simulation, First Edition.
Reihaneh Kardehi Moghaddam and Mostafa Rabbani.

Published 2025 by John Wiley & Sons, Inc.

Sliding mode control is especially useful for systems that are highly nonlinear and have uncertain dynamics. It can also be used for systems with external disturbances that need to be rejected. This technique is widely used in aerospace, automotive, and robotics applications.

1.2 Advantages and Limitations of Sliding Mode Controllers

Sliding mode control is a variable structure controller that was first introduced by Vadim Utkin [1]. Sliding mode control is a robust control method that is widely used to control and stabilize nonlinear systems. Its most important features are simple design and resistance to uncertainty in the model, i.e. parametric uncertainties and unmodeled dynamics, as well as rejection of external disturbances. Uncertainties have destructive effects on nonlinear systems and strongly affect the design of closed-loop control system [2]. The SMC design is based on the definition of an arbitrary linear slip surface in order to guarantee the optimal performance of the closed-loop control system and asymptotic stability based on the concept of the Lyapunov stability theorem. As SMC does not result in the convergence of system states to zero in finite time, it is developed to terminal sliding mode control (TSMC) in order to achieve the finite-time convergence. In this method, a nonlinear sliding surface is used instead of the linear one applied in conventional SMC methods. Nonlinear sliding surfaces in TSMC have improved the transient-mode performance and, compared to conventional SMC, TSMC provides features such as fast convergence, finite-time stability, increased accuracy of steady-state response, and in addition enhanced robustness [3]. However, the conventional TSMC faces two major problems that need to be solved: the singularity problem and the chattering phenomenon. To solve the first problem in [4], a switching method between nonlinear sliding surface in TSMC and linear sliding surface in SMC is proposed. Also, in [5], the method of transferring the system path, to a specific area in order to avoid singularity, is used. However, in these methods, the singularity problem is solved indirectly. Therefore, a nonsingular terminal sliding mode control (NTSMC) method has been introduced in [6], which directly solves the singularity problem by proposing a nonsingular nonlinear sliding surface.

Regarding the second problem, in techniques based on sliding mode control, when the system states are close to the sliding level, the existence of the discontinuous term in the control law causes chattering phenomenon. Chattering is an undesirable phenomenon as the vibration of the control signal causes energy loss, and, on the other hand, due to its high frequency, it may stimulate the high-frequency modes of the system which could potentially lead to instability. Also, chattering can cause wear in mechanical parts and electromechanical systems, and possible damage to various system components. To overcome this problem, various techniques are used in different articles such as high-order sliding mode control [7, 8], fuzzy sliding mode control [9–11], and using neural networks [12, 13].

Although, as stated, the supplementary topics of the sliding controllers, which are provided to solve the limitations of this controller, are very extensive, in this book, an attempt has been made to the basic part of these supplementary topics, i.e. chattering removal methods, terminal, fuzzy, adaptive, and super-twisting SMC. It is also combined with feedback linearization and back-stepping and at the end of each chapter a MATLAB programming is presented on the inverted pendulum to investigate the results.

1.3 An Overview of Book Contents

This book contains 12 chapters. In Chapter 1, sliding mode controllers and their challenges are introduced in general. In Chapter 2, the basic issues and definitions of classic sliding mode controllers and how to program and simulate these controllers in MATLAB have been discussed. Chapter 3 deals with the terminal sliding controllers that use the nonlinear sliding surface to increase the convergence speed. In addition, its complementary topics, including fast and nonsingular sliding mode controllers, have been introduced and simulated. Chapter 4 deals with the dynamic sliding controllers, and Chapter 5 introduces the fuzzy sliding controllers for the uncertain dynamic estimation of the system with fuzzy structure and its sliding control. In Chapter 6, adaptive sliding controllers with indeterminate parameters in system dynamics are presented along with parameter adaptation rules and MATLAB programming. Chapter 7 includes super-twisting sliding mode controllers. In Chapter 8, the combination of sliding and back-stepping control, which is of great interest due to its advantages such as the ability to design subsystems step by step, has been introduced and simulated. In Chapter 9, the phenomenon of chattering and the effective solutions to reduce drawbacks have been reviewed along with providing suitable examples and analytical tables of the results. Chapter 10 indicates the supplementary topics of using sliding controllers with feedback linearization. In Chapter 11, the design and stability analysis of fixed-time sliding mode controllers is discussed, and, finally, the last chapter is related to design and MATLAB programming of event-triggered sliding mode controllers.

The purpose of the authors of this book is to introduce basic techniques of SMC improvement to students and researchers who are interested in the design and simulation of sliding mode controllers and the control of nonlinear systems, quickly and easily. We hope that this book can be an effective guide for researchers in this field.

References

1 Utkin, V. (1977). Variable structure systems with sliding modes. *IEEE Transactions on Automatic Control* 22 (2): 212–222.

2 Slotine, J.J.E. and Li, W. (1991). *Applied Nonlinear Control*. Englewood Cliffs, NJ: Prentice Hall.

3 Zhihong, M., Paplinski, A.P., and Wu, H.R. (1994). A robust MIMO terminal sliding mode control scheme for rigid robotic manipulators. *IEEE Transactions on Automatic Control* 39 (12): 2464–2469.

4 Zhihong, M. and Yu, X.H. (1997). Terminal sliding mode control of MIMO linear systems. *IEEE Transactions on Circuits and Systems I: Fundamental Theory and Applications* 44 (11): 1065–1070.

5 Wu, Y., Yu, X., and Man, Z. (1998). Terminal sliding mode control design for uncertain dynamic systems. *Systems & Control Letters* 34 (5): 281–287.

6 Feng, Y., Yu, X., and Man, Z. (2002). Non-singular terminal sliding mode control of rigid manipulators. *Automatica* 38 (12): 2159–2167.

7 Bartolini, G., Ferrara, A., and Usai, E. (1998). Chattering avoidance by second-order sliding mode control. *IEEE Transactions on Automatic Control* 43 (2): 241–246.

8 Levant, A. (2005). Homogeneity approach to high-order sliding mode design. *Automatica* 41 (5): 823–830.

9 Nekoukar, V. and Erfanian, A. (2011). Adaptive fuzzy terminal sliding mode control for a class of MIMO uncertain nonlinear systems. *Fuzzy Sets and Systems* 179 (1): 34–49.

10 Cerman, O. and Hušek, P. (2012). Adaptive fuzzy sliding mode control for electro-hydraulic servo mechanism. *Expert Systems with Applications* 39 (11): 10269–10277.

11 Razzaghian, A. and Moghaddam, R.K. (2016). Adaptive fuzzy sliding mode control for a model-scaled unmanned helicopter. *Journal of Fuzzy Set Valued Analysis* 2016 (3): 286–302.

12 Zhao, L. and Jia, Y. (2016). Neural network-based distributed adaptive attitude synchronization control of spacecraft formation under modified fast terminal sliding mode. *Neurocomputing* 171: 230–241.

13 Razzaghian, A. and Moghaddam, R.K. (2018). Robust adaptive neural network control of miniature unmanned helicopter. *26th Iranian Conference on Electrical Engineering (ICEE 2018)*, 801–805, Mashhad, Iran (May 2018). IEEE.

2

Classic Sliding Mode Controller

2.1 Introduction

Sliding mode control (SMC) is a nonlinear control method in control systems that utilizes discontinuous control to modify the dynamics of a nonlinear system. The purpose of SMC is to force the system to slide along a predetermined surface that ensures desirable system behavior. This control method is a type of variable structure control designed to move states toward a specific sliding level. It can be viewed as a sub-branch of state feedback controllers. The history of feedback control can be traced back to ancient civilizations, where it was used in various applications such as level control, water clocks, and pneumatic/hydraulic systems. From the 17th century onward, control systems like temperature control in furnaces, mechanical control of mills, and steam engine regulators were developed. However, in the 19th century, feedback controllers were found to be prone to instability and oscillations, leading to system depreciation. However, they do not account for uncertainty, which prompted the development of advanced control techniques, including SMC.

2.2 Variable Structure Controllers

As mentioned earlier, the sliding mode controller is a type of variable structure controllers. Before explaining how to design sliding mode controller, it is necessary to point out that by using a proper variable structure and determining an appropriate rule to switch between these structures, a stable combination can be achieved. This stability can even result from the combination of several unstable systems. It can be clarified, considering the following example.

Methods of Developing Sliding Mode Controllers: Design and MATLAB Simulation, First Edition.
Reihaneh Kardehi Moghaddam and Mostafa Rabbani.

Published 2025 by John Wiley & Sons, Inc.

Consider the following system:

$$\begin{bmatrix} \dot{x}_2 \\ \dot{x}_2 \end{bmatrix} = \begin{bmatrix} 0 & 1 \\ 1 & 2 \end{bmatrix} \begin{bmatrix} x_1 \\ x_2 \end{bmatrix} + \begin{bmatrix} 0 \\ 1 \end{bmatrix} u \tag{2.1}$$

And let's assume that two controlling suggestions $u = 2x_1$ and $u = -3x_1$ are considered. By placing u and calculating the eigen values of the resulting systems, it is concluded that using $u = 2x_1$, the equilibrium point of the system is an unstable focal point, and considering the second suggestion, this equilibrium point is an unstable saddle point [1]. The system trajectory of these two unstable structures is shown in Figure 2.1.

Considering a sliding surface S and switching law $k(S, x_1)$, an asymptotically stable structure of Figure 2.2 will be obtained which leads to a stable structure:

$$S: \quad c_1x_1 + x_2 = 0 \;\; c_1 \geq 0 \tag{2.2}$$

$$u = kx_1; \quad k(S,x_1) = \begin{cases} 2 & Sx_1 < 0 \\ -3 & Sx_1 > 0 \end{cases} \tag{2.3}$$

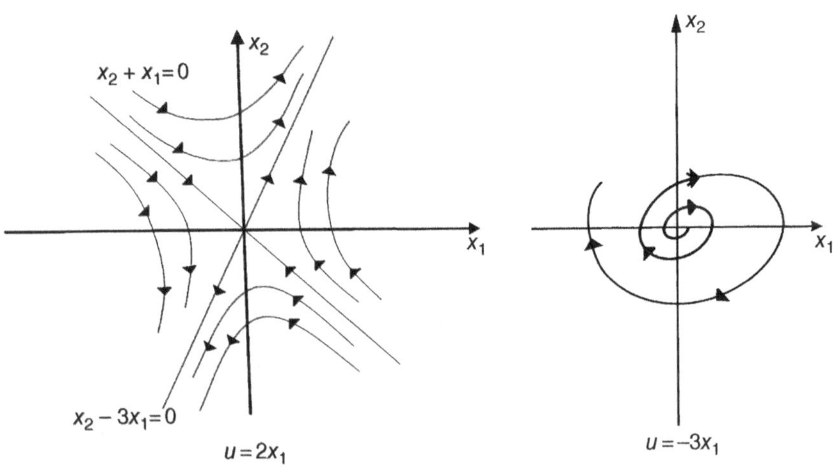

Figure 2.1 System trajectories for $u = 2x_1$ and $u = -3x_1$.

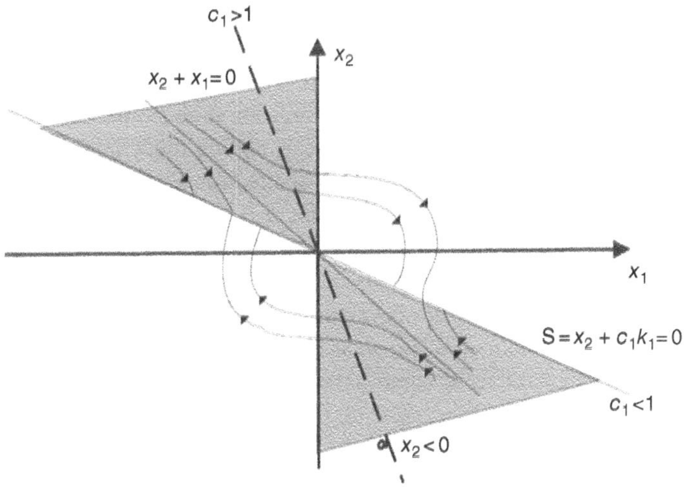

Figure 2.2 System trajectory after applying variable structure controller.

2.2.1 Industrial Variable Structure Controllers

2.2.1.1 Relays

One of the most famous variable structure controllers is relay, in which the controller includes an independent controlling law, and switches based on the sliding surface value which is usually a function of system states:

$$u_i = \begin{cases} u_i^{+} & S_i(x) > 0 \\ u_i^{-} & S_i(x) < 0 \end{cases} \tag{2.4}$$
$$S = [S_1, \ldots, S_m]T$$

2.2.1.2 Feedback Control with Switching Gains

In this structure, controller is a function of states and switches based on controlling gain values:

$$u = \varphi(x)x(t); \varphi = \left[\varphi_{ij}\right] \in R^{m \times n} \qquad \varphi_{ij} = \begin{cases} \alpha_{ij} & S_i(x)x_j > 0 \\ \beta_{ij} & S_i(x)x_j < 0 \end{cases} \tag{2.5}$$

2.2.1.3 Sliding Mode Controllers

SMC is a widely used variable structure control technique that drives the system state to a specified sliding surface and maintains it there through high-frequency switching. In this chapter this controller is discussed.

2.3 Classic Sliding Mode Controllers

Sliding mode controllers were firstly investigated in the early 1950s by Emalyanov and other Russian researchers for SISO systems, and, for such simple systems, sliding mode control is performed by trial and error. Sliding mode control did not progress much and was not welcomed until 1970 due to the lack of a specific method for designing the sliding surface. In 1970, a lot of attention was paid to this control method due to its specific structure; as a result, the research and development of sliding mode control design methods in both theoretical and practical fields became faster [2]. Since 1990, the development of the sliding mode and its integration with other control methods such as back-stepping control, adaptive control, and fuzzy control as well as its combination with population-based intelligent methods that increase the accuracy of the system have been carried out [3–9].

As mentioned above, uncertainties may arise in the modeling of most systems due to possible unmodeled dynamics and nonlinear effects neglected in the linearization of system equations, as well as disturbance and noise. These uncertainties can also appear due to disturbances in the system caused by changing the values of system parameters. Uncertainty in the model is divided into two general categories, which will be mentioned below.

Structured or parametric uncertainties which are rooted in *measurement limitations and errors* during the initial experiments. They are a kind of inherent uncertainty of process-based models that remains even if all input data were error-free. The second group contains unstructured uncertainties which can be due to the purposeful simplification of the system dynamics and is related to the inaccuracy in the order of the system or estimation error, for example, when friction is considered linear in modeling.

Inaccuracy in modeling can have adverse effects on nonlinear systems. Therefore, they should be explicitly considered in any practical design. To control nondeterministic systems, it is necessary to design a controller to be resistant in the presence of these uncertainties. Such a controller can be designed from robust, adaptive, and fuzzy control methods. Another method to control these types of systems is to use sliding mode control theory, which is one of the nonlinear and robust control methods for controlling uncertain systems. In this section, this control method and its developed types are explained.

Consider the following nonlinear system:

$$x^{(n)} = f(x,t) + g(x,t)u(t) + d(t) \tag{2.6}$$

where $f(x,t)$ and $g(x,t)$ are Lipshitz functions which may be uncertain but have definite limits (indeterminate or unmeasurable parameters, unmodeled dynamics). These functions are nonlinear and limited by a continuous and bounded function.

$d(t)$ is an indeterminate function and is considered as a disturbance or noise input with a predetermined upper and lower bounds. x is the state vector of the system, n is the order of the system, and u is the controlling signal, and it is assumed that the system is described in a canonical form.

2.4 Controller Design

The SMC is designed in two steps:

1) Choosing a stable hyperplane in the state space where the movement must be limited, and it is called a sliding surface. If the path of the system state is placed on this hyperplane, it should show appropriate behavior, that is, to bring the initial state of the system to the desired state.
2) The controller must have attraction properties and bring the paths left from the initial states to the sliding surface in a limited time, and also keep the states on the sliding surface. This is done by applying an appropriate control law [8–10].

Figure 2.3 shows the operation of the sliding mode controller.

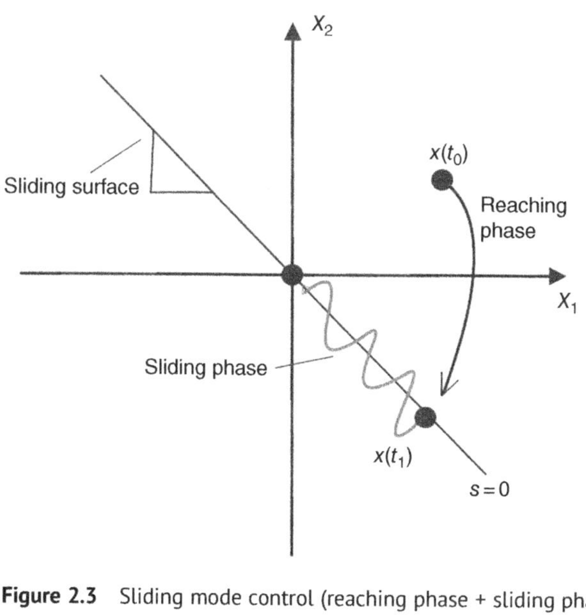

Figure 2.3 Sliding mode control (reaching phase + sliding phase).

2.5 Sliding Mode Controller Design

2.5.1 Reaching Phase

In Eq. (2.6), the control input has appeared in the nth derivative of x, and therefore the relative degree of the system will be equal to n. In the standard sliding mode control, first, for ease of design, weighted sum of the error and its derivatives are denoted by a new variable called the sliding surface. A general form for the sliding surface is defined in Eq. (2.7):

$$s = \left(\frac{d}{dt} + \lambda\right)^{n-1} \tilde{x} \tag{2.7}$$

in which λ is a positive weighting coefficient, $\tilde{x}$ denotes state variable errors, and x is vector of system states.

$$\tilde{x} = x - x_d \tag{2.8}$$

In the following, the control aim is to stabilize the system's states and their convergence to zero. For this purpose, with the formation of sliding dynamics, we have:

$$\dot{s}(t) = \frac{d^n(\tilde{x})}{dt^n} + \cdots + \lambda^n \tilde{x} = \tilde{x}^{(n)} + \cdots + \lambda^n \tilde{x} \tag{2.9}$$

Placing (2.8) in (2.9) yields:

$$\dot{s}(t) = x^{(n)} - x_d{}^{(n)} + \cdots + \lambda^n (x - x_d) = f(x,t) + g(x,t)u + d(t) - x_d{}^{(n)} + \cdots + \lambda^n \tilde{x} \tag{2.10}$$

Considering the above relationship, the control input appears in the first derivative of the sliding variable, and therefore the relative degree of this system is equal to 1. So, by changing the variable and defining the sliding variable, problems with relative degree n become one with relative degree 1. Now, to control this first-order system, Lyapunov's stability theory is used. For this purpose, the Lyapunov candidate function is considered as follows:

$$V = \frac{1}{2} s^2(t) \tag{2.11}$$

V is a certain positive function and is a measure of the distance from the sliding surface, and the control goal is the stability of the system, or in other words, reducing the distance from the sliding surface. According to Lyapunov stability theory [1], if the derivative of Lyapunov function in accordance with system 2.6 is negative definite, then the system will be asymptotically stable. To ensure the finite-time convergence to the sliding surface, it is considered to establish relation (2.12):

$$\dot{V} = \frac{d}{dt}\left(\frac{1}{2}s^2(t)\right) = s(t)\dot{s}(t) \leq -\eta|s(t)| \tag{2.12}$$

where η is a positive constant. Integrating both sides of the above equation results as:

$$\Rightarrow \int_{s(0)}^{0} \frac{s(t)}{|s(t)|} ds(t) \leq \int_{0}^{tr} -\eta dt \tag{2.13}$$

$$\Rightarrow \begin{cases} t_r \leq \dfrac{s(0)}{\eta}, \forall s(t) > 0 \\ t_r \leq \dfrac{-s(0)}{\eta}, \forall s(t) < 0 \end{cases}$$

It should be noted that in the above relation, the upper limit of the integral ($s(t_r)$) is equal to zero because the value of s on the sliding surface is equal to zero. Therefore, the convergence time of the sliding variable can be calculated from the following equation, which indicates reaching the sliding surface in a limited time:

$$t_r \leq \frac{|s(0)|}{\eta} \tag{2.14}$$

As it is indicated in this relation, the convergence time of sliding variable is limited and can be adjusted by changing the value of the η parameter.

2.5.2 Sliding Phase

When the states of system reach the sliding surface in a limited time, the controller must be designed to ensure that the system remains on the sliding surface without the presence of uncertainty, that is, the stability of the state variables must be checked; This controller is known as the nominal controller and is denoted by $\hat{u}$. In fact, in the nonlinear system with uncertainty, it is enough to estimate the uncertainty with a nominal value and then for a certain nominal system, it is found in such a way that $\dot{s} = 0$ (it does not change and the states remain on the sliding surface).

2.6 Theoretical Foundations of Sliding Mode Control

In this section, some fundamental theorems of SMC design are proposed.

Theorem 2.1

Consider the nonlinear system (2.6) with proposed Lyapunov function (2.11). A sufficient condition for the stability of sliding mode controller is $s\dot{s} < 0$ (that means $\frac{dV}{dt} < 0$) in a neighborhood of the surface given by $s = 0$ [1].

Theorem 2.2

To guarantee achieving sliding manifold in a finite time, $\frac{dV}{dt}$ must be more strongly bounded away from zero. To ensure that the sliding mode is entered in finite time the following condition is considered [1]:

$$\frac{dV}{dt} \leq -\mu\left(\sqrt{V}\right)^{\alpha}; \mu > 0; 0 \leq \alpha \leq 1$$

2.7 Designing Sliding Mode Controller for Second-Order Uncertain Systems

2.7.1 System Dynamical Model

Consider the following system:

$$\begin{cases} \dot{x}_1 = x_2 \\ \dot{x}_2 = f + gu + d \end{cases} \tag{2.15}$$

which can also be described as follows:

$$\ddot{x} = f + gu + d \tag{2.16}$$

As mentioned earlier, f and g are not known but have predefined limits and d is an indeterminate function and is considered as disturbance or noise input to the system.

$$f_{min} \leq f \leq f_{max}; g_{min} \leq g \leq g_{max}; |d| \leq D \tag{2.17}$$

Usually, for the nominal values of f and g, arithmetic and geometric mean values are used, respectively.

$$\hat{f} = \frac{f_{min} + f_{max}}{2}, \hat{g} = \sqrt{g_{min}g_{max}} \Rightarrow \sqrt{\frac{g_{min}}{g_{max}}} \leq \frac{g}{\hat{g}} \leq \sqrt{\frac{g_{max}}{g_{min}}} = (\beta > 0)$$
$$\Rightarrow 0 < \beta^{-1} < 1 \tag{2.18}$$

2.7.2 Sliding Surface Determination

To find the nominal controller from the sliding phase, $\dot{s}$ is calculated and considered to be zero.

$$s = \left(\frac{d}{dt} + \lambda\right)^{n-1} \tilde{x} \quad \overset{n=2}{\longrightarrow} \quad s = \dot{\tilde{x}} + \lambda\tilde{x} \Rightarrow \dot{s} = \ddot{\tilde{x}} + \lambda\dot{\tilde{x}} \tag{2.19}$$

2.7.3 Designing Nominal Controller

In the nominal conditions and with the assumption of ignoring the disturbances and uncertainties, the functions are placed with their nominal values, and then the nominal controller is obtained in such a way that the condition $\dot{s} = 0$ is fulfilled. This condition guarantees the maintenance of the system trajectories on the sliding surface.

$$\begin{aligned} &\Rightarrow \dot{s} = \ddot{x} - \ddot{x}_d + \lambda\dot{\tilde{x}} = 0 \\ &\Rightarrow \dot{s} = \hat{f} + \hat{g}\hat{u} - \ddot{x}_d + \lambda\dot{\tilde{x}} = 0 \\ &\Rightarrow \hat{u} = \frac{1}{\hat{g}}\left(-\hat{f} + \ddot{x}_d - \lambda\dot{\tilde{x}}\right) \end{aligned} \tag{2.20}$$

By applying the controller, it can only be guaranteed that the system states will follow the desired signal in nominal conditions and without uncertainty.

2.7.4 Designing the Variable Structure

In the presence of uncertainty and disturbance, a sentence with a variable structure must be added to the nominal controller, which shows the absorption property of the sliding surface and the survival of the states on the surface; for this purpose, the **sign** function is used. Therefore, the sliding mode controller will be as follows:

$$u = \hat{u} - \frac{k}{\hat{g}}\,\text{sign}(s) \Rightarrow u = \frac{1}{\hat{g}}\left(-\hat{f} + \ddot{x}_d - \lambda\dot{\tilde{x}} - k\,\text{sign}(s)\right) \tag{2.21}$$

$$\text{sign}(s) = \begin{cases} +1, \forall s > 0 \\ -1, \forall s > 0 \end{cases}$$

The value of k is chosen to be large enough to counteract modeling uncertainties. However, increasing k also leads to an increase in the controller amplitude. Therefore, an optimal value for k should be determined that effectively compensates for the model's uncertainties while maintaining the lowest possible controller magnitude. For this purpose and by establishing the condition of reaching

(2.12) and considering the system (2.15) in non-nominal conditions, the value of k is obtained as follows:

$$\dot{V} = \frac{d}{dt}\left(\frac{1}{2}s^2(t)\right) = s(t)\dot{s}(t) \leq -\eta|s(t)| \tag{2.22}$$

$$\Rightarrow s\left(f + gu + d - \ddot{x}_d + \lambda\dot{\tilde{x}}\right) \leq -\eta|s| \rightarrow$$

Substituting u (2.20) yields:

$$\Rightarrow s\left(f + g\left(\hat{u} - \frac{k}{\hat{g}}\,\text{sign}(s)\right) + d - \ddot{x}_d + \lambda\dot{\tilde{x}}\right) \leq -\eta|s| \rightarrow$$

$$\Rightarrow s\left(f + g\left(\frac{1}{\hat{g}}\left(-\hat{f} + \ddot{x}_d - \lambda\dot{\tilde{x}}\right) - \frac{k}{\hat{g}}\,\text{sign}(s)\right) + d - \ddot{x}_d + \lambda\dot{\tilde{x}}\right) \leq -\eta|s|$$

$$\Rightarrow s\left(f - g\hat{g}^{-1}\hat{f} + g\hat{g}^{-1}\ddot{x}_d - \lambda g\hat{g}^{-1}\dot{\tilde{x}} - kg\hat{g}^{-1}\,\text{sign}(s) + d - \ddot{x}_d + \lambda\dot{\tilde{x}}\right) \leq -\eta|s|$$

$$\Rightarrow s\left(f - g\hat{g}^{-1}\hat{f} - kg\hat{g}^{-1}\,\text{sign}(s) + d + (g\hat{g}^{-1} - 1)\ddot{x}_d - (g\hat{g}^{-1} - 1)\lambda\dot{\tilde{x}}\right) \leq -\eta|s|$$

Adding and subtracting $\hat{f}$ yields:

$$\Rightarrow s\left(f - \hat{f} + (g\hat{g}^{-1} - 1)\underbrace{\left(-\hat{f} + \ddot{x}_d - \lambda\dot{\tilde{x}}\right)}_{\hat{u}} - kg\hat{g}^{-1}\,\text{sign}(s) + d\right) \leq -\eta|s|$$

$$\Rightarrow s\left(f - \hat{f} + (g\hat{g}^{-1} - 1)\underbrace{\left(-\hat{f} + \ddot{x}_d - \lambda\dot{\tilde{x}}\right)}_{\hat{u}} - kg\hat{g}^{-1}\,\text{sign}(s) + d\right) \leq -\eta|s|$$

$$\Rightarrow s\left(f - \hat{f} + (g\hat{g}^{-1} - 1)\hat{u}\right) - ksg\hat{g}^{-1}\text{sign}(s) + sd \leq -\eta|s|$$

$$\Rightarrow |s|\left|f - \hat{f} + (g\hat{g}^{-1} - 1)\hat{u}\right| - kg\hat{g}^{-1}|s| + |d||s| \leq -\eta|s| \rightarrow$$

Using triangular inequality results in:

$$\Rightarrow |s|\underbrace{\left|f - \hat{f}\right|}_{F} + \left|g\hat{g}^{-1} - 1\right||\hat{u}||s| - kg\hat{g}^{-1}|s| + D|s| \leq -\eta|s|$$

$$\Rightarrow \begin{cases} s > 0 \overset{s=|s|}{\rightarrow} F + D + \left|g\hat{g}^{-1} - 1\right||\hat{u}| \leq k - \eta \\ s < 0 \overset{s=-|s|}{\rightarrow} -F - D - \left|g\hat{g}^{-1} - 1\right||\hat{u}| \geq k - \eta \end{cases} \tag{2.23}$$

$$g \Rightarrow k \geq \beta(F + \eta + D) + (\beta - 1)|\hat{u}|$$

$$\Rightarrow \forall\beta = 1, k \geq F + \eta + D$$

Therefore, the appropriate value of k is obtained, which will lead to the design of a resistant and stabilizing controller. In this section, the design process of classical sliding mode controller with linear sliding surface was investigated [11–15]. In the next section, the proposed controller is simulated using MATLAB.

2.8 Classical Sliding Mode Controller Design in MATLAB

Example 2.1 Design of SMC for a second-order nonlinear system.
To clarify the controlling method, consider the following uncertain nonlinear system:

$$\ddot{x} + a(t)\dot{x}^2 \cos 3x = u$$
$$1 \le a(t) \le 2$$

The goal is to design a suitable sliding controller such that the state variable x follows the path $x_d = \sin(\pi t/2)$. The following procedure is used to design the sliding mode controller. First, the nominal value of the uncertain parameter a, denoted by $\hat{a}$, is calculated as follows:

$$\hat{a} = 1.5 \Rightarrow \hat{f} = 1.5\dot{x}^2 \cos 3x$$

Then, taking into account that the system is of the second order, based on relation (2.7), the sliding surface is suggested as follows:

$$s = \left(\frac{d}{dt} + \lambda\right)\tilde{x} = \dot{\tilde{x}} + \lambda\tilde{x}$$

Assuming the nominal condition (absence of disturbance and parametric uncertainty), the nominal part of the sliding controller ($\hat{u}$) is calculated so that ($\dot{s} = 0$) be satisfied.

$$\dot{s} = \ddot{\tilde{x}} + \lambda\dot{\tilde{x}} = \ddot{x} - \ddot{x}_d + \lambda\dot{\tilde{x}} = \left[u - a(t)\dot{x}^2 \cos 3x\right] - \ddot{x}_d + \lambda\dot{\tilde{x}}$$
$$\dot{s} = 0; a = a = 1.5$$
$$\hat{u} = \hat{f} + \ddot{x}_d - \lambda\dot{\tilde{x}} = \left[1.5\dot{x}^2 \cos 3x\right] + \ddot{x}_d - \lambda\dot{\tilde{x}}$$

Adding the sign term to the nominal controller, the k value is obtained as follows. In this section, the uncertainties and disturbances, if any, are considered.

$$s\dot{s} \le -\mu|s| \Rightarrow s\left\{\left[\left[u - a(t)\dot{x}^2 \cos 3x\right] - \ddot{x}_d + \lambda\dot{\tilde{x}}\right\}\right.$$
$$\le -\mu|s| \xRightarrow{u = \hat{u} - k\,\mathrm{sgn}(s)} s\left\{\left[\hat{u} - K\,\mathrm{sgn}(s) - a(t)\dot{x}^2 \cos 3x\right] - \ddot{x}_d + \lambda\dot{\tilde{x}}\right\} \le -\mu|s|$$

$$\xrightarrow{\hat{u}=-\hat{a}+\ddot{x}_d-\lambda\dot{\tilde{x}}} s\left\{\left[\hat{a}\dot{x}^2\cos 3x+\ddot{x}_d-\lambda\dot{\tilde{x}}-k\,\mathrm{sgn}(s)-a(t)\dot{x}^2\cos 3x\right]-\ddot{x}_d+\lambda\dot{\tilde{x}}\right\}\le -\mu|s|$$

$$\Rightarrow\ s\{\left[\hat{a}\dot{x}^2\cos 3x-k\,\mathrm{sgn}(s)-a\dot{x}^2\cos 3x\right]\}\le-\mu|s|\Rightarrow s\left[(-a+\hat{a})\dot{x}^2\cos 3x\right]$$
$$-ks\,\mathrm{sgn}\,s=s(\hat{F}-F)-k|s|\le-\mu|s|$$
$$\Rightarrow s(\hat{F}-F)\le(k-\mu)|s|$$
$$s(\hat{F}-F)\le(k-\mu)|s|\Rightarrow\begin{cases}\{(\hat{F}-F)\le(k-\mu)\ \text{if}\ s>0\\-(\hat{F}-F)\ge(k-\mu)\ \text{if}\ s<0\end{cases}$$
$$\Rightarrow k-\mu\ge\left|(\hat{F}-F)\right|=|(\hat{a}-a)|\dot{x}^2|\cos 3x|=0.5\dot{x}^2|\cos 3x|$$

With arbitrary selection of $\mu = 0.1$, the acceptable value for coefficient k is calculated as follows based on the above inequality.

$$k = 0.1 + 0.5\dot{x}^2|\cos 3x|$$

$$\begin{aligned}u &= \hat{u} - \frac{k}{g}\,\mathrm{sgn}(s); k = \left(0.5\dot{x}^2|\cos 3x| + .1\right); \ddot{x}_d\\ &= \left(\frac{-\pi^2}{4}\right)\sin(\pi t/2); \dot{\tilde{x}} = \dot{x} - \left(\frac{\pi}{2}\right)\cos(\pi t/2)\\ u &= 1.5\dot{x}^2\cos 3x + \ddot{x}_d - 20\dot{\tilde{x}} - \left(0.5\dot{x}^2|\cos 3x| + 0.1\right)\mathrm{sgn}\left(\dot{\tilde{x}} + 20\tilde{x}\right)\end{aligned} \tag{2.24}$$

$$\begin{aligned}\dot{x}_1 &= x_2\\ \dot{x}_2 &= -a(t)\dot{x}^2\cos 3x + u\end{aligned} \tag{2.25}$$

To simulate the results, MATLAB functions are written as follows. The system's dynamic is described in function below:

```
    function dx = classic_sliding(t,x)

    dx = zeros(2,1);
 % desired trajectory and its derivatives
    xd = sin(pi*t/2);
    xd_d = (pi/2)*(cos(pi*t/2));
    xd_dd = -(pi^2/4)*(sin(pi*t/2));
 %controller structure equation (2.24)
    s = x(2)-xd_d +20*(x(1)-xd);
    U = 1.5*(x(2)^2)*cos(3*x(1)) +xd_dd  -20*(x(2)-xd_d)-
(0.5*(x(2)^2)*abs(cos(3*x(1)))+0.1)*sign(s);
 %system's dynamic (equation 2.25)
 % a is an unknown constant between 1 and 2
     a = (abs(sin(t))+1);
     dx(1) = x(2);
     dx(2) = -a*(x(2)^2)*cos(3*x(1))+U;
```

The execution mfile is as follows:

```
%PLOT STATES AND TRACKING ERROR
    [t,x] = ode23(@classic_sliding,[0 20],[0.5 0]);
    figure(1)
    plot( t,x(:,1),'b--',t,sin(pi*t/2),'r','linewidth',2)
    xlabel('t(s)')
    legend('x1','x desired')
        x1 = x(:,1);x2 = x(:,2);
        [m,n] = size(t);s = zeros(m,1);U = zeros(m,1);
    for i = 1:m
        xd = sin(pi*t(i)/2);
        xd_d = (pi/2)*(cos(pi*t(i)/2));
        xd_dd = -(pi^2/4)*(sin(pi*t(i)/2));
        s(i) = x2(i)-xd_d+20*(x1(i)-xd);
   U(i) = 1.5*((x2(i))^2)*cos(3*x1(i)) +xd_dd -20*(x2(i)-
xd_d)-(0.5*((x2(i))^2)*abs(cos(3*x1(i)))+0.1)*sign(s(i));
    end
    figure(2)
      plot( t,U,'b')
      xlabel('t(s)')
      ylabel('controller')
      figure(3)
    error = x(:,1)- sin(pi*t/2);
      xlabel('t(s)')
      plot( t,error,'k.')
      ylabel('error')
```

Figures 2.4–2.6 show the controlling signals and also systems states.

As shown in Figures 2.4–2.6, after 16 seconds, the tracking error is almost equal to zero. Figure 2.4 shows the unwanted fluctuations of the controller, especially in the 18–20 seconds interval, which is known as chattering.

The above system can also be simulated as in Figure 2.7, where Simulink MATLAB functions are defined as follows:

```
function U = control(x1,x2,t)
 xd = sin(pi*t/2);
 xd_d = (pi/2)*(cos(pi*t/2));
 xd_dd = -(pi^2/4)*(sin(pi*t/2));

 s = x2-xd_d +20*(x1-xd);
 U = 1.5*(x2^2)*cos(3*x1) +xd_dd-20*(x2-xd_d)-
   (0.5*(x2^2)*abs(cos(3*x1))+0.1)*sign(s);

 function [dx1,dx2]= system(U,t,x1,x2)
 a = (abs(sin(t))+1);
```

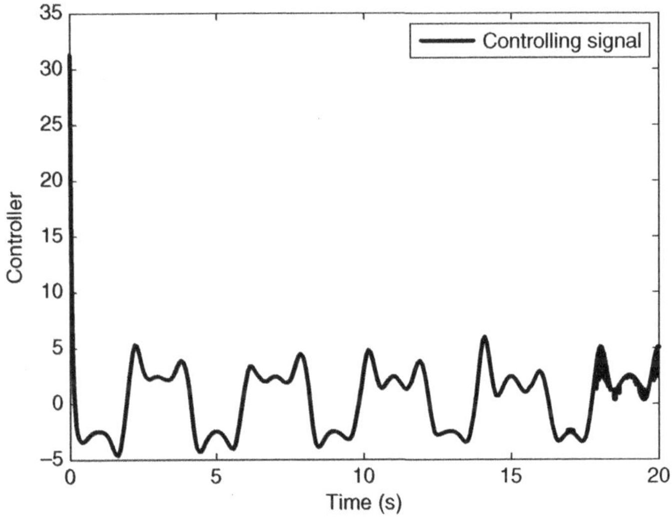

Figure 2.4 Controlling signal.

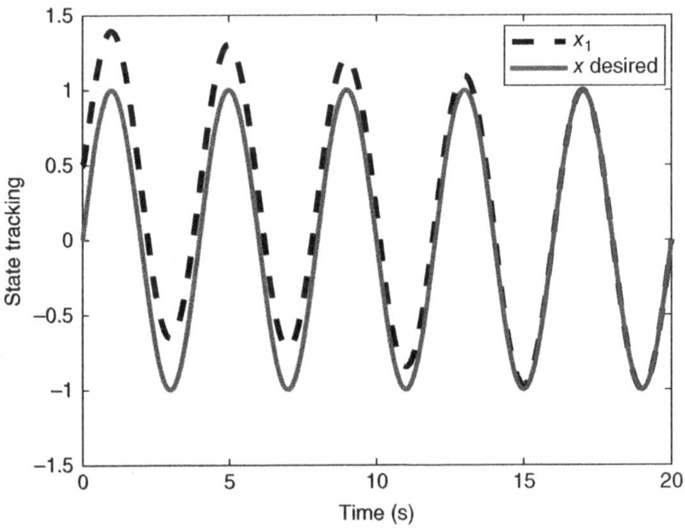

Figure 2.5 State tracking.

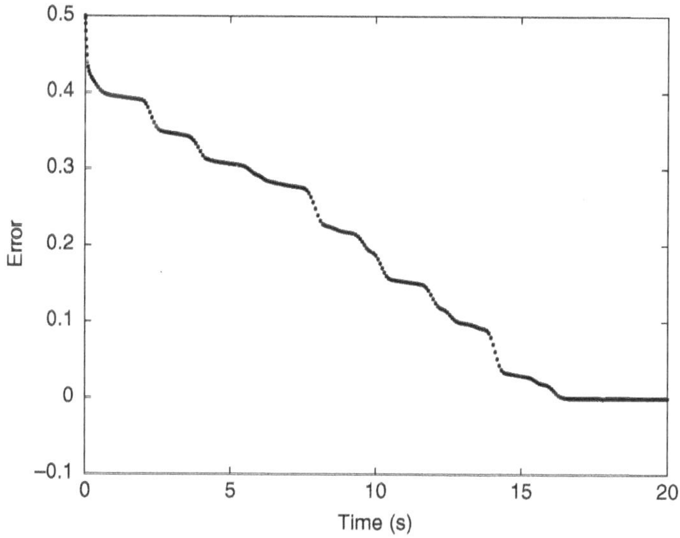

Figure 2.6 Tracking error.

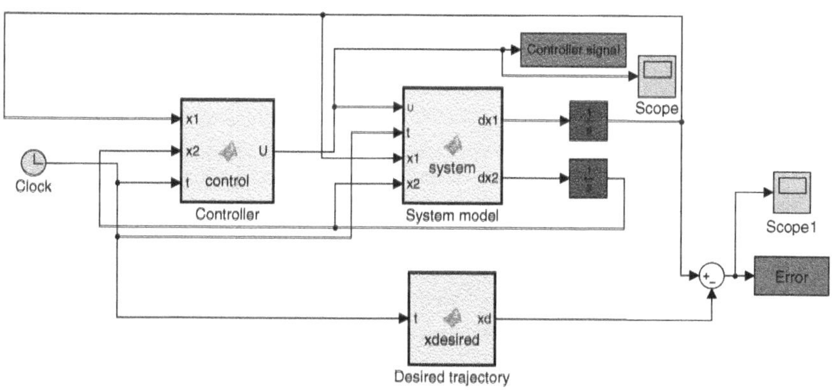

Figure 2.7 Systems block diagram in SIMULINK.

```
dx1 = x2;
dx2 = -a*(x2^2)*cos(3*x1)+U;
```

```
function xd = xdesired(t)
xd = sin(pi*t/2);
```

The tracking error and controlling signal are proposed in Figures 2.8 and 2.9.

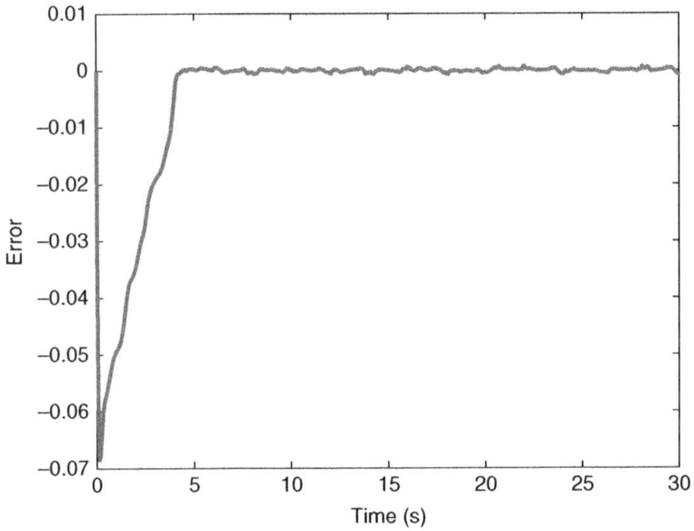

Figure 2.8 Tracking error.

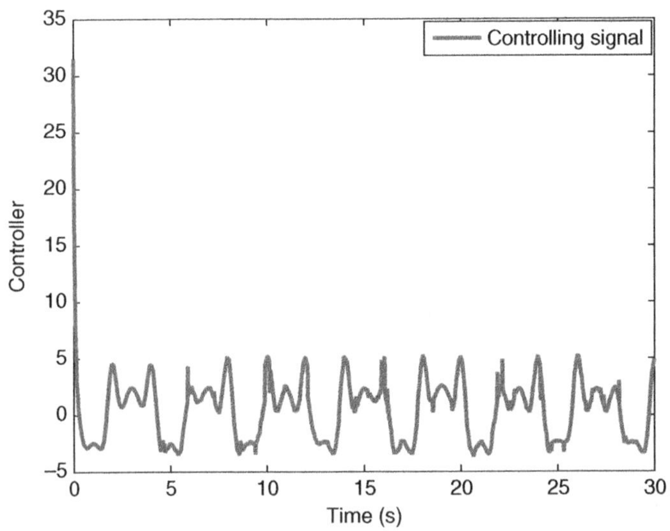

Figure 2.9 Controlling signal.

Example 2.2 Designing sliding mode control for an inverted pendulum
Consider the inverted pendulum structure as follows (Figure 2.10):

$$
\begin{aligned}
\dot{x_1} &= x_2 \\
\dot{x_2} &= f(x) + g(x)u + d \\
f(x) &= \frac{g\ \sin x_1 - mlx_2^2\ \cos x_1\ \sin x_1/(m_c + m)}{l(\frac{4}{3} - m\ \cos^2 x_1/(m_c + m))}; \\
g &= \frac{9.8\ \text{m}}{s^2}, m_c = 1, m = 0.1\ \text{kg}, l = 0.5\ \text{m} \\
g(x) &= \frac{\cos x_1/(m_c + m)}{l(\frac{4}{3} - m\ \cos^2 x_1/(m_c + m))}
\end{aligned}
\tag{2.26}
$$

Replacing f and g in the controller proposed in (2.21), we obtain the controller as follows:

$$
u = \frac{1}{\hat{g}}\left(-\hat{f} + \ddot{x}_d - \lambda\dot{\tilde{x}} - k\ \text{sign}(s)\right) \tag{2.27}
$$

To design the sliding surface we consider (2.7) and the SMC is considered to be as (2.27). k is supposed to be 5.

At first the system's dynamic and controlling structure is proposed in SMC_INVPENDULUM MATLAB function.

```
function dx=SMC_INVPENDULUM(t,x)
    dx = zeros(2,1);
  % desired trajectory and its derivatives
    xd = sin(pi*t/2);
    xd_d = (pi/2)*(cos(pi*t/2));
    xd_dd = -(pi^2/4)*(sin(pi*t/2));
    %PARAMETER VALUES
```

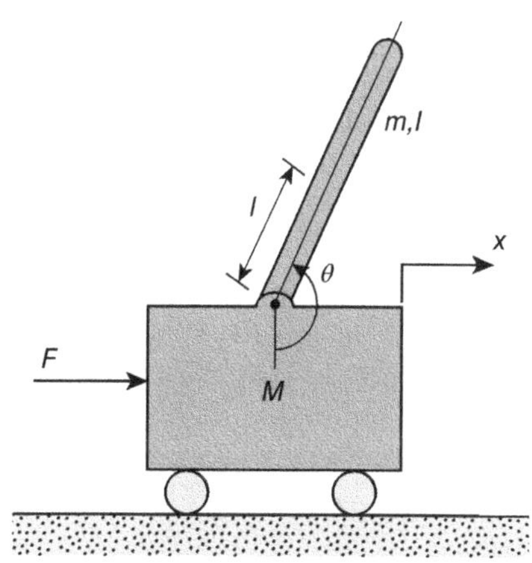

Figure 2.10 Inverted pendulum.

```
    g = 9.8;mc = 1.0;m = 0.1;l = 0.5;landa = 1;
    h = l*(4/3-m*(cos(x(1)))^2/(mc+m));
    fx = g*sin(x(1))-m*l*(x(2))^2*cos(x(1))*sin(x(1))/(mc
+m);
    fx = fx/h;
    gx = cos(x(1))/(mc+m);
    gx = gx/h;
    %controller structure equation (3.8) sliding surface
(3.6)
    q = 3;p = 5;k = 5;gxhat = gx; fxhat = fx;
    s = x(2)-xd_d +landa*(x(1)-xd);
    U = -inv(gxhat)*(fxhat+landa*(x(2)-xd_d)-xd_dd+k*sign
(s));

    dx(1) = x(2);
    dx(2) = fx+gx*U;
```

To execute the controller, the following MATLAB function is considered.

```
%PLOT STATES AND TRACKING ERROR

[t,x] = ode23(@SMC_INVPENDULUM,[0 20],[0.5 0]);
 xd = sin(pi*t/2);
 xd_d = (pi/2)*(cos(pi*t/2));
figure(1)
subplot(211)
plot( t,x(:,1),':r',t,xd,'k','linewidth',2)
xlabel('t(s)')
ylabel('x1')
legend('x1','x1 desired')
subplot(212)
plot( t,x(:,2),':r',t,xd_d,'k','linewidth',2)
xlabel('t(s)')
ylabel('x2')
legend('x2','x2 desired')
x1 = x(:,1);x2 = x(:,2);
    [m,n] = size(t);s = zeros(m,1);U = zeros(m,1);
for i = 1:m
    xd = sin(pi*t(i)/2);
    xd_d = (pi/2)*(cos(pi*t(i)/2));
    xd_dd = -(pi^2/4)*(sin(pi*t(i)/2));
     g = 9.8;mc = 1.0;m = 0.1;l = 0.5;landa = 1;
     h = l*(4/3-m*(cos(x1(i)))^2/(mc+m));
     fx = g*sin(x1(i))-m*l*(x2(i))^2*cos(x1(i))*sin(x1(i))/
(mc+m);
```

```
      fx = fx/h;
      gx = cos(x1(i))/(mc+m);
      gx = gx/h;
      q = 3;p = 5;k = 5;gxhat = gx; fxhat = fx;
     s(i) = (x2(i)-xd_d) +landa*(x1(i)-xd);
     U(i) = -inv(gxhat)*(fxhat-xd_dd+landa*(x2(i)-xd_d)
+k*sign(s(i)));

end
figure(2)
subplot(211)
plot( t,U,'b')
xlabel('t(s)')
ylabel('U(n)')
subplot(212)
error1 = x(:,1)- sin(pi*t/2);
error2 = x(:,2)- (pi/2)*(cos(pi*t/2));
xlabel('t(s)')
```

Figures 2.11 and 2.12 show the tracking process and controlling signal. As it is illustrated, system's states follow the desired signal, and unwanted chattering is

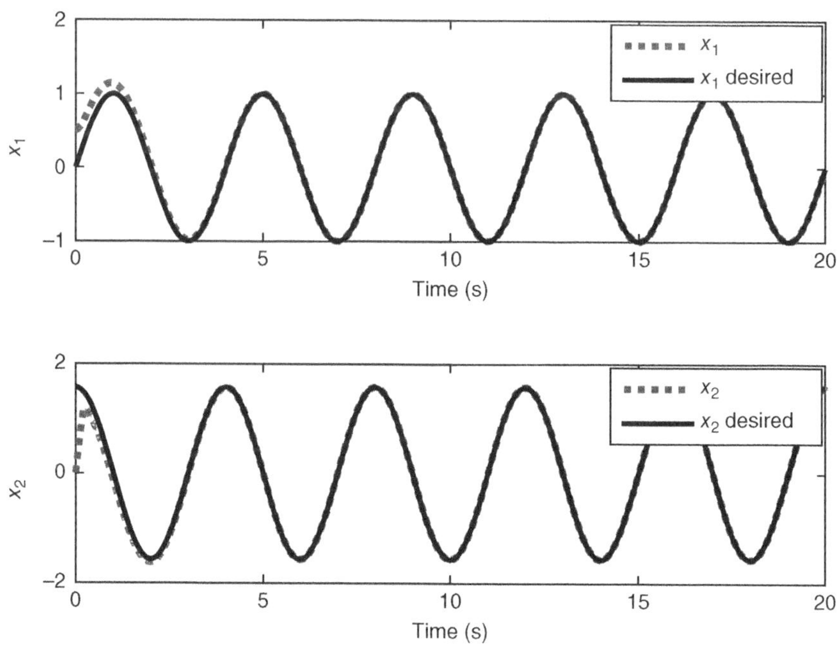

Figure 2.11 Inverted pendulum position tracking using classic SMC.

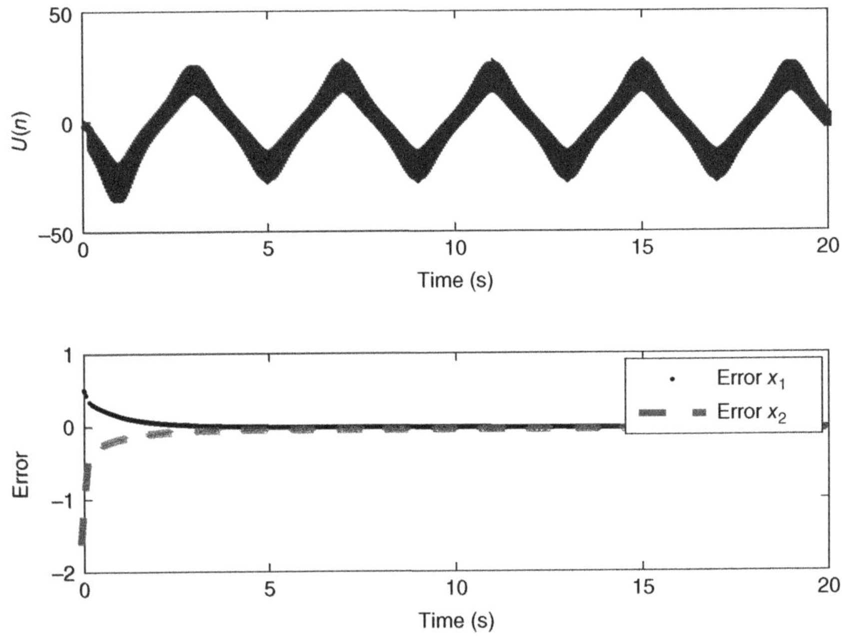

Figure 2.12 Inverted pendulum controller and tracking error using SMC.

observed in Figure 2.12 which is an undesirable consequence of using classic sliding mode controllers.

References

1 Slotine, J.-J.E. and Li, W. (1991). *Applied Nonlinear Control.* Prentice Hall.
2 Edwards, C., Colet, E.F., and Fridman, L. (2006). *Advances in Variable Structure and Sliding Mode Control.* Springer.
3 Razzaghian, A. and Moghaddam, R.K. (2016). Adaptive fuzzy sliding mode control for a model-scaled unmanned helicopter. *Journal of Fuzzy Set Valued Analysis* 2016 (3): 286–302.
4 Razzaghian, A., Moghaddam, R.K., and Pariz, N. (2021). Fractional-order nonsingular terminal sliding mode control via a disturbance observer for a class of nonlinear systems with mismatched disturbances. *Journal of Vibration and Control* 27 (1–2): 140–151.
5 Razzaghian, A., Moghaddam, R.K., and Pariz, N. (2021). Disturbance observer-based fractional-order nonlinear sliding mode control for a class of fractional-order

systems with matched and mismatched disturbances. *International Journal of Dynamics and Control.* 9 (2): 671–678.

6 Razzaghian, A., Kardehi Moghaddam, R., and Pariz, N. (2020). Adaptive neural network conformable fractional-order nonsingular terminal sliding mode control for a class of second-order nonlinear systems. *IETE Journal of Research* 1–10.

7 Baghernia, P., Moghaddam, R.K., and Kobravi, H. (2015). Cancer sliding mode control considering to chaotic manners of system. *Journal of Medical Imaging and Health Informatics* 5 (3): 448–457.

8 Moghaddam, R.K. and Mohamadie, N.K. (2023). Terminal nonsingular sliding mode control of wave energy converters for maximum power absorption. *Ocean Engineering* 281 (1): 114473.

9 Shamsabadi, M. and Moghaddam, R.K. (2020). Underwater robot trajectory control using nonsingular terminal sliding mode controller. *International Journal of Smart Electrical Engineering* 10 (3): 117–125.

10 Shtessel, Y., Edwards, C., Fridman, L., and Levant, A. (2014). *Sliding Mode Control and Observation.* Springer.

11 Imahe, O. (2015). *From Differential Equations Through Lyapunov Functions to Sliding Mode Control (v1.0).* Afe Babalola University.

12 Bartoszewicz, A. (2011). *Sliding Mode Control,* Printed in India.

14 Sabanovic, A., Fridman, L.M., and Spurgeon, S. (2004). *Variable Structure Systems from principles to implementation.* London: The Institution of Engineering and Technology.

15 Shtessel, Y., Edwards, C., Fridman, L., and Levant, A. (2014). *Sliding Mode Control and Observation.* New York: Springer Science+Business Media.

3

Terminal Sliding Mode Controller

3.1 Introduction

The terminal sliding mode controller (TSMC) was first presented in 1990 by Venkataraman and Gulati, followed by Zack in 1998 [1]. One of the disadvantages of the classical sliding mode controller is the low convergence rate, which is due to the use of the linear sliding surface function. To improve the convergence rate, it is necessary to greatly increase the controller's gain. The controller with a high gain value is not suitable in practice due to saturation problems. Therefore, the TSMC, which is based on a nonlinear sliding surface function and effectively improves convergence speed, has been introduced and utilized in various applications.

3.2 Terminal Sliding Mode Controller Design

Consider a single input affine nonlinear system as follows:

$$\begin{aligned} \dot{x}_1 &= x_2 \\ \dot{x}_2 &= f(x,t) + g(x,t)u(t) + d(t) \end{aligned} \tag{3.1}$$

where $\boldsymbol{x} = [\ x_1\ x_2] \in R^2$ is the system-states vector. x_d is the desired trajectory, $\tilde{x}_1 = x_1 - x_d$ is the tracking error, and $\tilde{x}_2 = x_2 - \dot{x}_d$. f and g are nonlinear affine functions, and $d \le |D|$ denotes the bounded disturbance. Although f and g can include parametric uncertainties, their related nominal values $\hat{f}$ and $\hat{g}$ are known. The limited variations around the nominal values are considered as follows:

$$\left|f - \hat{f}\right| \le F, |g/\hat{g}| \le \beta^{-1} \tag{3.2}$$

Methods of Developing Sliding Mode Controllers: Design and MATLAB Simulation, First Edition.
Reihaneh Kardehi Moghaddam and Mostafa Rabbani.

Published 2025 by John Wiley & Sons, Inc.

To design a TSMC for the above system, the following sliding surface is considered:

$$s = \tilde{x}_2 + \lambda(\tilde{x}_1)^{\frac{q}{p}}, \lambda > 0 \tag{3.3}$$

In the above relation, q and p are odd and $p > q$. It is clear that for even q values the sliding surface should be defined as $s = \tilde{x}_2 + \lambda(|\tilde{x}_1|)^{\frac{q}{p}} \operatorname{sign}(\tilde{x}_1)$ to obtain an admissible controlling signal [2]. In sliding phase, states are placed on the sliding surface, so s is equal to zero and the following relation is considered:

$$\dot{\tilde{x}}_1 = -\lambda(\tilde{x}_1)^{\frac{q}{p}} \operatorname{sign}(\tilde{x}_1) \tag{3.4}$$

Definition 3.1 Reaching Time
The time duration taken for system states to start from $\boldsymbol{x}(0)$ and reach the sliding surface s. It is denoted by (t_r) [3].

Definition 3.2 Sliding Time
The time duration taken for system states to start from $\boldsymbol{x}(t_r)$ and reach the equilibrium point x_e. It is denoted by (t_s).

Solving the equation above, the sliding time (t_s) is calculated as follows [4]:

$$\frac{d\tilde{x}_1}{dt} = -\lambda\tilde{x}_1^{\frac{q}{p}} \Rightarrow \int_{\tilde{x}_1(t_r)}^{0} \tilde{x}_1^{-\frac{q}{p}} d\tilde{x} = \int_0^{t_s} -\lambda dt \Rightarrow \left.\frac{\tilde{x}_1^{1-\frac{q}{p}}}{1-\frac{q}{p}}\right|_{\tilde{x}(t_r)}^{0} = -\lambda t\Big|_0^{t_s}$$

$$t_s = \frac{\left|\tilde{x}_1(t_r)^{1-\frac{q}{p}}\right|}{\lambda\left(1-\frac{q}{p}\right)} = \frac{p}{\lambda(p-q)}\left|\tilde{x}_1(t_r)^{1-\frac{q}{p}}\right| \tag{3.5}$$

The relation above confirms that the systems-tracking error slides to zero in a finite sliding time. Terminal sliding mode control and conventional sliding mode control can be compared not only based on their sliding phase time but also in terms of their reaching time. Both criteria are essential for evaluating the performance and efficiency of these control techniques in regulating a system's response. TSMC is an extension of SMC, designed to achieve faster convergence and improved control performance. It employs a nonlinear sliding surface that provides finite-time convergence, meaning that the system's states reach the sliding surface within a predefined, finite-time interval. This feature is particularly beneficial for applications requiring rapid and precise control actions. In contrast, conventional SMC can exhibit slower convergence, especially when the system's initial conditions are far from the sliding surface. Although it also provides robustness against modeling uncertainties and external disturbances, its reaching time might be longer compared to TSMC. Therefore, when comparing TSMC and SMC, it is crucial to consider both sliding time and reaching time to assess the overall control system performance accurately. Depending on the specific application and its requirements, either control method may be more suitable. For system (3.1), the reaching time t_r for SMC and TSMC is calculated using the reaching condition inequality, $s\dot{s} \leq -\eta|s|$, as follows:

$$\text{if } s \geq 0 \rightarrow \int_{s=s(0)}^{s=s(t_r)} ds \leq \int_{t=0}^{t=t_r} -\eta\, dt; \qquad s(t_r) - s(0) \leq -\eta\ t_r \rightarrow t_r \leq \frac{s(0)}{\eta} \tag{3.6}$$

$$\text{if } s < 0 \rightarrow \int_{s=s(0)}^{s=s(t_r)} ds > \int_{t=0}^{t=t_r} \eta\, dt; \qquad s(t_r) - s(0) > \eta\ t_r \rightarrow t_r \leq \frac{-s(0)}{\eta}$$

In the above relations $s(t_r) = 0$ which is concluded from the fact that in t_r system trajectories reach the sliding surface. Relation (3.6) results in t_r boundedness in TSMC and SMC. Considering the above relation, it is concluded that with the assumptions of $p = q$, the classic sliding mode controller is obtained. So, in classic SMC, the denominator of (3.5) is zero, and consequently, the convergence time of the states to the equilibrium point is infinite.

It is obvious that t_s is adjustable by defining (p, q, λ) [3]. The significant characteristic of TSMC is its fast convergence near the equilibrium point $((|x_1| < 1))$. Considering the sliding surface (3.3), the nominal part of TSMC is defined as (3.8).

$$s = \tilde{x}_2 + \lambda\left(\tilde{x}_1^{\frac{q}{p}}\right) \Rightarrow \dot{s} = \dot{\tilde{x}}_2 + \lambda\frac{q}{p}\dot{\tilde{x}}_1(\tilde{x}_1)^{\frac{q}{p}-1} \tag{3.7}$$

Remark 3.1

Without loss of generality the term $\tilde{x}_1^{\frac{q}{p}}$ can be replaced by $|\tilde{x}_1|^{\frac{q}{p}}\operatorname{sign}(\tilde{x}_1)$ to avoid imaginary s values.

Having substituted (3.1) in (3.7) and differentiated s, the nominal controller is calculated as follows:

$$\dot{s} = \hat{f} + \hat{g}\hat{u} - \ddot{x}_d + \lambda\frac{q}{p}\dot{\tilde{x}}_1(\tilde{x}_1)^{\frac{q}{p}-1} \Rightarrow \hat{u} = -\hat{g}^{-1}\left(\hat{f} + \lambda\frac{q}{p}\tilde{x}_2(\tilde{x}_1)^{\frac{q}{p}-1} - \ddot{x}_d\right) \tag{3.8}$$

Considering uncertainty, the variable structure part is added to nominal controller as follows:

$$u = \hat{u} - k\hat{g}^{-1}\operatorname{sign}(s)$$

$$\Rightarrow u = -\hat{g}^{-1}\left(\hat{f} + \lambda\frac{q}{p}\tilde{x}_2(\tilde{x}_1)^{\frac{q}{p}-1} - \ddot{x}_d + k\operatorname{sign}(s)\right) \tag{3.9}$$

Theorem 3.1

The TSMC, proposed in (3.9), stabilizes system (3.1).

Proof: Consider the Lyapunov function $V = \frac{1}{2}s^2$

To prove the stability the following reaching condition should be satisfied:

$$\dot{V} = \frac{d}{dt}\left(\frac{1}{2}s^2(t)\right) = s(t)\dot{s}(t) \leq -\eta|s(t)| \tag{3.10}$$

Substituting controller proposed in (3.9) in inequality (3.10) yields:

$$s\left(f+g\left\{-\hat{g}^{-1}\left(\hat{f}+\lambda\frac{q}{p}\tilde{x}_2(\tilde{x}_1)^{\frac{q}{p}-1}-\ddot{x}_d+k\,\mathrm{sign}(s)\right)\right\}+d-\ddot{x}_d+\lambda\frac{q}{p}\dot{\tilde{x}}_1(\tilde{x}_1)^{\frac{q}{p}-1}\right)$$
$$\le-\eta|s|$$

$$s\left(f-g\hat{g}^{-1}\hat{f}-g\hat{g}^{-1}\lambda\frac{q}{p}\tilde{x}_2(\tilde{x}_1)^{\frac{q}{p}-1}-g\hat{g}^{-1}k\,\mathrm{sign}(s)+g\hat{g}^{-1}\ddot{x}_d-\ddot{x}_d+d+\lambda\frac{q}{p}\tilde{x}_2(\tilde{x}_1)^{\frac{q}{p}-1}\right)$$
$$\le-\eta|s|$$

Addition and subtraction of $\hat{f}$ leads to:

$$s\left(f-\hat{f}+\hat{f}-g\hat{g}^{-1}\hat{f}-g\hat{g}^{-1}\lambda\frac{q}{p}\tilde{x}_2(\tilde{x}_1)^{\frac{q}{p}-1}-g\hat{g}^{-1}k\,\mathrm{sign}(s)+g\hat{g}^{-1}\ddot{x}_d-\ddot{x}_d+d+\lambda\frac{q}{p}\tilde{x}_2(\tilde{x}_1)^{\frac{q}{p}-1}\right)$$
$$\le-\eta|s|$$

$$\Rightarrow s\underbrace{\left(f-\hat{f}\right)}_{F}+s(1-g\hat{g}^{-1})\hat{f}+s(1-g\hat{g}^{-1})\lambda\frac{q}{p}\tilde{x}_2(\tilde{x}_1)^{\frac{q}{p}-1}$$
$$-(1-g\hat{g}^{-1})\ddot{x}_d-g\hat{g}^{-1}k\,\mathrm{sign}(s)+d)\le-\eta|s|$$

$$\Rightarrow|s|\left(F+\left(1-\beta^{-1}\right)\left(\hat{f}+\lambda\frac{q}{p}\tilde{x}_2(\tilde{x}_1)^{\frac{q}{p}-1}-\ddot{x}_d\right)-\beta^{-1}k\,\mathrm{sign}(s)+D\right)\le-\eta|s|$$

$$\Rightarrow k\ge\beta(F+\eta+D)+(\beta-1)|\hat{u}|\Rightarrow\forall\beta=1,k\ge F+\eta+D \tag{3.11}$$

So, using a proper k, the stability condition will be satisfied.

3.3 Nonsingular Terminal Sliding Mode Controller

Considering (3.9) it is clear that TSMC contains $\lambda(q/p)\tilde{x}_2(\tilde{x}_1)^{\frac{q}{p}-1}$ term, and as $\frac{q}{p}-1<0$, in condition $\tilde{x}_1=0,\tilde{x}_2\neq0$, the singularity happens. In this situation the controller value is infinite and to solve the problem, a nonsingular terminal SMC (NTSMC) with the following sliding surface is proposed:

$$s=\tilde{x}_1+\lambda\tilde{x}_2^{\frac{p}{q}} \tag{3.12}$$

Remark 3.2
In special conditions, where even q is considered, without loss of generality the term $\tilde{x}_2^{\frac{p}{q}}$ can be replaced by $|\tilde{x}_2|^{\frac{p}{q}}\,\mathrm{sign}(\tilde{x}_2)$ to avoid imaginary s values.

In the sliding phase states are supposed to be on the sliding surface, so s is considered to be zero.

$$\tilde{x}_1 = -\lambda(\tilde{x}_2)^{\frac{p}{q}} \Rightarrow \tilde{x}_2 = \dot{\tilde{x}}_1 = -\frac{1}{\lambda^{\frac{q}{p}}}\tilde{x}_1^{\frac{q}{p}} \tag{3.13}$$

Solving (3.13), the sliding time is calculated as follows:

$$\frac{d\tilde{x}_1}{dt} = -\frac{1}{\lambda^{\frac{q}{p}}}\tilde{x}_1^{\frac{q}{p}} \Rightarrow \int_{x_1(t_r)}^{0} \tilde{x}_1^{-\frac{q}{p}} d\tilde{x}_1 = \int_0^{t_s} -\frac{1}{\lambda^{\frac{q}{p}}} dt \Rightarrow \left.\frac{\tilde{x}_1^{1-\frac{q}{p}}}{1-\frac{q}{p}}\right|_{\tilde{x}_1(t_r)}^{0} = \left. -\frac{1}{\lambda^{\frac{q}{p}}} t \right|_0^{t_s}$$

$$t_s = \frac{\left|\tilde{x}_1(t_r)^{1-\frac{q}{p}}\right|}{\frac{1}{\lambda^{\frac{q}{p}}}\left(1-\frac{q}{p}\right)} = \frac{\lambda^{-\frac{q}{p}}p}{p-q}\left|\tilde{x}_1(t_r)^{1-\frac{q}{p}}\right| \tag{3.14}$$

From (3.14), it is concluded that the system states converge to the desired trajectory in a finite time.

To define the nominal controller the relation $\dot{s} = 0$ is considered, where the sliding surface is defined in (3.13).

$$\begin{aligned} s = \tilde{x}_1 + \lambda \tilde{x}_2^{\frac{p}{q}} &\Rightarrow \dot{s} = \dot{\tilde{x}}_1 + \lambda\frac{p}{q}\dot{\tilde{x}}_2(\tilde{x}_2)^{\frac{p}{q}-1} \\ &\Rightarrow \dot{s} = \tilde{x}_2 + \lambda\frac{p}{q}\left(\hat{f} + \hat{g}\hat{u} - \ddot{x}_d\right)(\tilde{x}_2)^{\frac{p}{q}-1} = 0 \end{aligned} \tag{3.15}$$

$$\Rightarrow \hat{u} = -\hat{g}^{-1}\left(\hat{f} - \ddot{x}_d + \frac{q}{\lambda p}\tilde{x}_2^{2-\frac{p}{q}}\right) \tag{3.16}$$

In the presence of uncertainties and disturbances, it is necessary to add a variable structure term to the nominal controller. This term guarantees the state maintenance on sliding surfaces. The proper parameter values should be defined in a way that they guarantee the reachability condition as defined in Theorem 3.2.

$$\begin{aligned} &u = \hat{u} - k\hat{g}^{-1}\mathrm{sign}(s) \\ &\Rightarrow u = -\hat{g}^{-1}\left(\hat{f} - \ddot{x}_d + \frac{q}{\lambda p}\tilde{x}_2^{2-\frac{p}{q}} + k\,\mathrm{sign}(s)\right) \end{aligned} \tag{3.17}$$

This control input contains a sentence $\tilde{x}_2^{2-\frac{p}{q}}$ and will not lead to singularity. To prove the stability, Theorem 3.2 will be discussed. In the sequel, for the purpose of simplifying the proofs and avoiding repeated calculations g is considered to be known ($g = \hat{g}$).

Theorem 3.2
The NTSMC, proposed in (3.17) stabilizes system (3.1).

Proof: Consider the Lyapunov function $V = \frac{1}{2}s^2$

To prove the stability, the following reaching condition is considered (inequality (3.10)):

$$\dot{V} = \frac{d}{dt}\left(\frac{1}{2}s^2(t)\right) = s(t)\dot{s}(t) \leq -\eta|s(t)|$$

Substituting the controller proposed in (3.17) in inequality (3.10) yields:

$$s\left(\tilde{x}_2 + \lambda\frac{p}{q}(\tilde{x}_2)^{\frac{p}{q}-1}\left[f + g\left\{-\hat{g}^{-1}\left(\hat{f} - \ddot{x}_d + \frac{q}{\lambda p}\tilde{x}_2{}^{2-\frac{p}{q}} + k\,\text{sign}(s)\right)\right\} + d - \ddot{x}_d\right)\right] \leq -\eta|s|$$

$$\overset{g=\hat{g}}{\longrightarrow} s\left(\tilde{x}_2 + \lambda\frac{p}{q}(\tilde{x}_2)^{\frac{p}{q}-1}\left[f - \hat{f} - \frac{q}{\lambda p}\tilde{x}_2{}^{2-\frac{p}{q}} - k\,\text{sign}(s)\} + d\right]\right) \leq -\eta|s|$$

$$k \geq D + F + \eta\left(\lambda\frac{p}{q}(x_2)^{\frac{p}{q}-1}\right)^{-1} \tag{3.18}$$

Here, the proper value for k which leads to systems stability is defined.

3.4 Fast Terminal Sliding Mode Controller

As illustrated in (3.3), in TSMC the convergence speed is inversely proportional to the distance from the equilibrium point. So in points far from the origin, TSMC performs low-speed convergence. To solve the problem, a fast terminal sliding-mode controller (FTSMC) which is a combination of classic and TSMCs is proposed. In fact, at the points near the equilibrium point, the controller takes advantage of fast convergence of TSMC and in points away from the equilibrium point, the classic SMC is applied. FTSMC is effectively used in systems that are far away from the equilibrium point, especially when the system is faced with extreme uncertainties and the TSMC has inappropriate performance.

The sliding surface of FTSMC is defined as follows:

$$s = \tilde{x}_2 + \alpha\tilde{x}_1 + \beta\tilde{x}_1{}^{\frac{q}{p}}\,(\alpha,\beta) > 0 \tag{3.19}$$

Remark 3.3

Without loss of generality the term $\tilde{x}_1{}^{\frac{q}{p}}$ can be replaced by $|\tilde{x}_1|^{\frac{q}{p}}\,\text{sign}(\tilde{x}_1)$ to avoid imaginary s values.

s is equal to zero during the sliding phase. Considering zero value for s, the following relation is established:

$$\tilde{x}_2 = -\alpha\tilde{x}_1 - \beta\tilde{x}_1{}^{\frac{q}{p}} \tag{3.20}$$

Relation (3.20) is rewritten as follows:

$$\dot{\tilde{x}}_1 = \underbrace{\alpha\tilde{x}_1}_{\text{LSMC}} \underbrace{- \beta\tilde{x}_1^{\frac{q}{p}}}_{\text{TSMC}} \tag{3.21}$$

The first part shows the linear sliding mode property and the second part indicates the terminal sliding mode performance. Solving the equation above, the sliding time is calculated as follows [3]:

$$t_s = \frac{p}{\alpha(p-q)}\left(\ln\left(\alpha(\tilde{x}_1(t_r))^{1-\frac{q}{p}} + \beta\right) - \ln(\beta)\right) \tag{3.22}$$

Theorem 3.3

The stabilizing FTSMC, which guarantees the limited sliding time for nonlinear system defined in (3.1), is as (3.23).

$$u = -\hat{g}^{-1}\left(\hat{f} - \ddot{x}_d + \alpha\tilde{x}_2 + \beta\frac{q}{p}\tilde{x}_2\tilde{x}_1^{\frac{q}{p}-1} + k\,\text{sign}(s)\right); \Rightarrow k \geq F + \eta + D \tag{3.23}$$

Proof: The nominal controller of fast terminal sliding mode is calculated using $\dot{s} = 0$, where s is defined in (3.19) and for simplicity g is considered to be known $(g = \hat{g})$.

$$\dot{s} = \hat{f} + g\hat{u} - \ddot{x}_d + \alpha\tilde{x}_2 + \beta\frac{q}{p}\tilde{x}_2\tilde{x}_1^{\frac{q}{p}-1} = 0$$

$$\Rightarrow \hat{u} = -g^{-1}\left(\hat{f} - \ddot{x}_d + \alpha\tilde{x}_2 + \beta\frac{q}{p}\tilde{x}_2\tilde{x}_1^{\frac{q}{p}-1}\right) \tag{3.24}$$

The general structure of FTSMC is as follows:

$$u = \hat{u} - kg^{-1}\,\text{sign}(s)$$

$$\Rightarrow u = -g^{-1}\left(\hat{f} - \ddot{x}_d + \alpha\tilde{x}_2 + \beta\frac{q}{p}\tilde{x}_2\tilde{x}_1^{\frac{q}{p}-1} + k\,\text{sign}(s)\right) \tag{3.25}$$

To prove the stability, the Lyapunov function $V = \frac{1}{2}s^2$ is considered. k should find in such a way that it guarantees the reachability condition (3.10).

Substituting (3.19) in (3.10) yields:

$$s\left(f + g\left\{-g^{-1}\left(\hat{f} - \ddot{x}_d + \alpha\tilde{x}_2 + \beta\frac{q}{p}\tilde{x}_2\tilde{x}_1^{\frac{q}{p}-1} + k\text{sign}(s)\right)\right\} + d - \ddot{x}_d + \alpha\tilde{x}_2 + \beta\frac{q}{p}\tilde{x}_2\tilde{x}_1^{\frac{q}{p}-1}\right) \leq -\eta|s|$$

$$\Rightarrow s\left(f - \hat{f} - \alpha\tilde{x}_2 - \beta\frac{q}{p}\tilde{x}_2\tilde{x}_1^{\frac{q}{p}-1} - k\,\text{sign}(s) + d + \alpha\tilde{x}_2 + \beta\frac{q}{p}\tilde{x}_2\tilde{x}_1^{\frac{q}{p}-1}\right) \leq -\eta|s|$$

$$\Rightarrow |s|(F - k + D) \leq -\eta|s|$$

$$\Rightarrow k \geq F + \eta + D \tag{3.26}$$

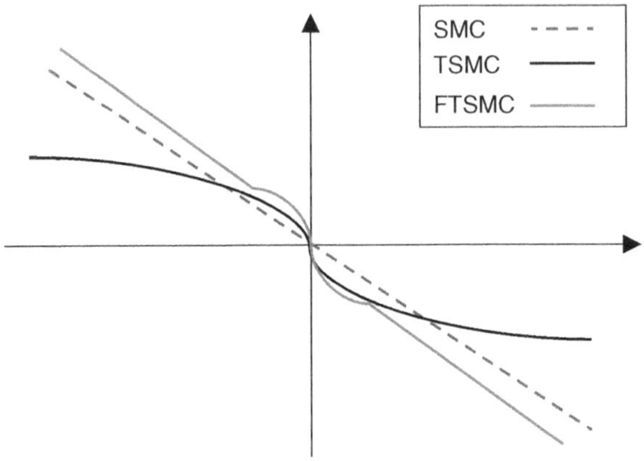

Figure 3.1 The sliding surface in linear sliding mode, terminal sliding mode, and fast terminal sliding mode controllers.

Figure 3.1 shows the sliding surface in classic sliding mode controller, TSMC, and FTSMC.

As shown in Figure 3.1, the slope of the sliding surface in the FTSMC is greater at distances far from the equilibrium point than in the TSMC, which is due to the addition of a linear term to the terminal sliding surface. This structure will increase the rate of convergence. It is obvious that an excessive increase in the linear term causes chattering in the control signal. Therefore, the coefficients of sliding surfaces must be chosen correctly. In addition, the parameters in the TSMC and the fast terminal sliding mode must be carefully selected to produce the maximum terminal attraction.

3.5 Nonsingular Fast Terminal Sliding Mode Control

One limitation of TSMC and FTSMC is the singularity phenomenon, which can be solved by proposing a sliding surface as a combination of terminal sliding or fast terminal sliding surface with that of nonsingular SMC. Therefore, in this book, the sliding surface of the NFTSMC is defined as follows [5–7]:

$$s = \tilde{x}_1 + \frac{1}{\alpha}\tilde{x}_1 + \frac{1}{\beta}\tilde{x}_2^{\frac{p}{q}} \tag{3.27}$$

where $(\alpha, \beta) > 0$,$1 < (p/q) < 2$ and $\{p, q\} \in N$ *odd*.

Assuming that the sliding mode controller is able to stabilize the system in limited time, we can consider sliding surface equal to zero during the sliding phase. Considering zero value for sliding surface, the following relation is established:

$$\tilde{x}_2 = \left(-\beta\left(\frac{1}{\alpha}+1\right)\right)^{\frac{q}{p}} \tilde{x}_1^{\frac{q}{p}} \tag{3.28}$$

Solving (3.28), the sliding time is extracted as follows:

$$\frac{d\tilde{x}_1}{dt} = \left(-\beta\left(\frac{1}{\alpha}+1\right)\right)^{\frac{q}{p}} \tilde{x}_1^{\frac{q}{p}} \Rightarrow \int_{\tilde{x}_1(t_r)}^{0} \tilde{x}_1^{-\frac{q}{p}} d\tilde{x}_1 = \int_0^{t_s} \left(-\beta\left(\frac{1}{\alpha}+1\right)\right)^{\frac{q}{p}} dt$$

$$\Rightarrow \left.\frac{\tilde{x}_1^{1-\frac{q}{p}}}{1-\frac{q}{p}}\right|_{\tilde{x}_1(t_r)}^{0} = \ldots$$

$$= \left(-\beta\left(\frac{1}{\alpha}+1\right)\right)^{\frac{q}{p}} t\Big|_0^{t_s} \Rightarrow t_s = \frac{\tilde{x}_1(t_r)^{1-\frac{q}{p}}}{\left(-\beta\left(\frac{1}{\alpha}+1\right)\right)^{\frac{q}{p}}\left(\frac{q}{p}-1\right)} \tag{3.29}$$

Theorem 3.4

The stabilizing NFTSMC for nonlinear system (3.1) is defined as follows (where for simplicity and avoiding repeated calculations, g is considered to be known):

$$u = -g^{-1}\left(\hat{f} + \frac{\left(1+\frac{1}{\alpha}\right)\tilde{x}_2^{2-\frac{p}{q}}}{\frac{1}{\beta}\frac{p}{q}} + k\,\text{sign}(s)\right) \tag{3.30}$$

Proof: The nominal controller is calculated using $\dot{s} = 0$:

$$s = \tilde{x}_1 + \frac{1}{\alpha}\tilde{x}_1 + \frac{1}{\beta}\tilde{x}_2^{\frac{p}{q}} \Rightarrow \dot{s} = \dot{\tilde{x}}_1 + \frac{1}{\alpha}\dot{\tilde{x}}_1 + \frac{1}{\beta}\frac{p}{q}\dot{\tilde{x}}_2\dot{\tilde{x}}_2^{\frac{p}{q}-1}$$

$$\dot{s} = \left(1+\frac{1}{\alpha}\right)\tilde{x}_2 + \frac{1}{\beta}\frac{p}{q}\left(\hat{f} + g\hat{u} - \ddot{x}_d\right)\tilde{x}_2^{\frac{p}{q}-1} = 0$$

$$\Rightarrow \hat{u} = -g^{-1}\left(\hat{f} - \ddot{x}_d + \frac{\left(1+\frac{1}{\alpha}\right)\tilde{x}_2^{2-\frac{p}{q}}}{\frac{1}{\beta}\frac{p}{q}}\right) \tag{3.31}$$

The NFTSMC is defined as follows:

$$u = \hat{u} - k\hat{g}^{-1}\text{sign}(s); g = \hat{g}$$

$$u = -g^{-1}\left(\hat{f} - \ddot{x}_d + \frac{\left(1+\frac{1}{\alpha}\right)\tilde{x}_2^{2-\frac{p}{q}}}{\frac{}{\beta}\frac{p}{q}} + k\,\text{sign}(s)\right) \tag{3.32}$$

To prove the stability, the Lyapunov function $V = \frac{1}{2}s^2$ is considered. k should find in such a way that it guarantees the condition $\dot{V} < 0$.

$$\dot{V} = \frac{d}{dt}\left(\frac{1}{2}s^2(t)\right) = s(t)\dot{s}(t) \le -\eta|s(t)| \tag{3.33}$$

$$\Rightarrow s\left(\left(1 + \frac{1}{\alpha}\right)\tilde{x}_2 + \frac{1}{\beta}\frac{p}{q}(f + gu - \ddot{x}_d + d)\tilde{x}_2^{\frac{p}{q}-1}\right) \le -\eta|s|$$

Substituting (3.32) in (3.33) yields

$$s\left(\left(1+\frac{1}{\alpha}\right)\tilde{x}_2+\frac{1p}{\beta q}\left(f-\ddot{x}_d+g\left\{-g^{-1}\left(\hat{f}-\ddot{x}_d+\frac{\left(1+\frac{1}{\alpha}\right)x_2^{2-\frac{p}{q}}}{\frac{1p}{\beta q}}+k\text{sign}(s)\right)\right\}+d\right)\tilde{x}_2^{\frac{p}{q}-1}\right)$$

$$\le -\eta|s|$$

$$\Rightarrow s\left(\underbrace{f-\hat{f}}_{F} - k\,\text{sign}(s) + D\right) \le -\frac{\eta}{\frac{1}{\beta}\frac{p}{q}\tilde{x}_2^{\frac{p}{q}-1}}|s| \tag{3.34}$$

$$\Rightarrow k \ge F + D + \beta\frac{q}{p}\left(\tilde{x}_2^{1-\frac{p}{q}}\right)\eta$$

So the above k guarantees the stability.

3.6 Integral Terminal Sliding Mode Control

In this part, an integral terminal sliding mode controller (INTSMC) is proposed which includes the advantages of integral SMC in reducing the chattering phenomena and TSMC in finite-time convergence simultaneously.

For system (3.1) the sliding surface is designed as follows [8]:

$$s = \tilde{x}_1 + \int_0^t \left(c_1\,\text{sign}(\tilde{x}_2)|\tilde{x}_2|^{\gamma_1} + c_2\,\text{sign}(\tilde{x}_1)|\tilde{x}_1|^{\gamma_2}\right)dt \tag{3.35}$$

where $\tilde{x}_1 = x_1 - x_d$ is the tracking error and $\tilde{x}_2 = x_2 - \dot{x}_d$. x_d is the desired trajectory. All constants above are positive and $1 < \gamma_1 < 2\,; \gamma_2 = \frac{\gamma_1}{2-\gamma_1}$ for the case where controlling aim is states' convergence to equilibrium point $\tilde{x}_1 = x_1$.

To guarantee finite-time convergence and chattering reduction, the INTSMC surface is considered as below:

$$\sigma = \dot{s} + \alpha s; \alpha > 0 \tag{3.36}$$

To obtain the nominal controlling part, it is assumed that $\dot{\sigma} = 0$. Combining (3.35), (3.36), and (3.1) the nominal controller is calculated.

$$\dot{\sigma} = 0 = \ddot{s} + \alpha\dot{s} = \ddot{x}_1 + c_1\gamma_1\ddot{x}_1|\tilde{x}_2|^{\gamma_1 - 1} + c_2\gamma_2\dot{\tilde{x}}_1|\tilde{x}_1|^{\gamma_2 - 1}$$
$$+ \alpha(c_1 \operatorname{sign}(\tilde{x}_2)|\tilde{x}_2|^{\gamma_1} + c_2 \operatorname{sign}(\tilde{x}_1)|\tilde{x}_1|^{\gamma_2})$$

$$= \ddot{x} - \ddot{x}_d + c_1\gamma_1(\ddot{x} - \ddot{x}_d)|\tilde{x}_2|^{\gamma_1 - 1} + c_2\gamma_2\dot{\tilde{x}}_1|\tilde{x}_1|^{\gamma_2 - 1}$$
$$+ \alpha\left(\dot{\tilde{x}}_1 + c_1 \operatorname{sign}(\tilde{x}_2)|\tilde{x}_2|^{\gamma_1} + c_2 \operatorname{sign}(\tilde{x}_1)|\tilde{x}_1|^{\gamma_2}\right)$$

$$= f + g\hat{u} - \ddot{x}_d + c_1\gamma_1(f + g\hat{u} - \ddot{x}_d)|\tilde{x}_2|^{\gamma_1 - 1} + c_2\gamma_2\dot{\tilde{x}}_1|\tilde{x}_1|^{\gamma_2 - 1}$$
$$+ \alpha(c_1 \operatorname{sign}(\tilde{x}_2)|\tilde{x}_2|^{\gamma_1} + c_2 \operatorname{sign}(\tilde{x}_1)|\tilde{x}_1|^{\gamma_2})$$

$$\hat{u} = -g^{-1}\left[-\ddot{x}_d + f + \left(\frac{\alpha}{1 + c_1\gamma_1|\tilde{x}_2|^{\gamma_1 - 1}}\left(\dot{\tilde{x}}_1 + c_1 \operatorname{sign}(\tilde{x}_2)|\tilde{x}_2|^{\gamma_1} + c_2 \operatorname{sign}(\tilde{x}_1)|\tilde{x}_1|^{\gamma_2}\right)\right.\right.$$
$$\left.\left. + \frac{1}{1 + c_1\gamma_1|\tilde{x}_2|^{\gamma_1 - 1}}\left(c_2\gamma_2\dot{\tilde{x}}_1|\tilde{x}_1|^{\gamma_2 - 1}\right)\right)\right] \tag{3.37}$$

The INTSMC is obtained by adding the switching term to the nominal part as follows:

$$u = \hat{u} - \frac{k}{g} \operatorname{sign}(\sigma) \tag{3.38}$$

Theorem 3.5
The INTSMC, proposed in (3.38), stabilizes system (3.1).

Proof: Consider the Lyapunov function $V = \frac{1}{2}\sigma^2$. To prove the stability of controller, k should be defined in such a way that it satisfies the reachability condition $\dot{V} = \frac{d}{dt}\left(\frac{1}{2}\sigma^2(t)\right) = \sigma(t)\dot{\sigma}(t) \leq -\eta|\sigma(t)|$.

$$\dot{V} = \sigma\dot{\sigma} = \sigma\Big[f + gu + d - \ddot{x}_d + c_1\gamma_1(f + gu + d - \ddot{x}_d)|\tilde{x}_2|^{\gamma_1 - 1} + c_2\gamma_2\dot{\tilde{x}}_1|\tilde{x}_1|^{\gamma_2 - 1}$$
$$+ \alpha(c_1 \operatorname{sign}(\tilde{x}_2)|\tilde{x}_2|^{\gamma_1} + c_2 \operatorname{sign}(\tilde{x}_1)|\tilde{x}_1|^{\gamma_2})\Big] =$$

$$\sigma\left[f + g\left(\hat{u} - \frac{k}{g}\operatorname{sign}(\sigma)\right) + d - \ddot{x}_d + c_1\gamma_1\left(f + g\left(\hat{u} - \frac{k}{g}\operatorname{sign}(\sigma)\right) + d - \ddot{x}_d\right)\right.$$
$$\left.|\tilde{x}_2|^{\gamma_1 - 1} + c_2\gamma_2\dot{\tilde{x}}_1|\tilde{x}_1|^{\gamma_2 - 1} + \alpha(c_1 \operatorname{sign}(\tilde{x}_2)|\tilde{x}_2|^{\gamma_1} + c_2 \operatorname{sign}(\tilde{x}_1)|\tilde{x}_1|^{\gamma_2})\right.$$

Replacing $\hat{u}$ by (3.37) yields:

$$\dot{V} = \sigma\dot{\sigma} = \sigma\left[(1 + c_1\gamma_1|\tilde{x}_2|^{\gamma_1 - 1})(d - k\operatorname{sign}(\sigma))\right]$$

choosing $k = \eta + D$ yields

$$\dot{V} = \left(\underbrace{1+\gamma_1|\tilde{x}_2|^{\gamma_1-1}}_{H}\right)(d\sigma-(\eta+D)\,|\sigma|) \leq H(D|\sigma|-(\eta+D)\,|\sigma|) \leq -H\eta|\sigma| < 0$$

So, the controller stabilizes system (3.1), where D is the upper bound of uncertainty. As it is mentioned in [9] without loss of generality the proposed method can be applied to n-order systems.

3.7 Extended INTSMC for *n*-Order Nonlinear Systems

The idea, which is discussed in previous part, can be extended for nth-order nonlinear systems with the following canonical structure [9, 11]:

$$\begin{aligned}
\dot{x}_1 &= x_2\\
\dot{x}_2 &= x_3\\
&\dots\\
\dot{x}_n &= f(x,t) + g(x,t)u(t) + d(t)
\end{aligned} \tag{3.39}$$

where $x = [\,x_1, \ldots, x_n] \in R^n$ is the system-states vector. x_d is the desired trajectory, $\tilde{x}_1 = x_1 - x_d$ is the tracking error, and $\tilde{x}_2 = x_2 - \dot{x}_d$. Disturbance d is considered to be bounded considering $d \leq |D|$.

For system (3.39) the ITSM sliding manifold is considered as follows:

$$s = \tilde{x}_1 + \int_0^t \left(\sum_{i=1}^{n} c_i \operatorname{sign}\left(\tilde{x}^{(n-i)}\right)\left|\tilde{x}^{(n-i)}\right|^{\gamma_i}\right) dt \tag{3.40}$$

where $\tilde{x}^{(k)} = x_k - x_d^{(k)}$ and $1 < \gamma_i < 2; \gamma_i = \dfrac{\gamma_{i-1}}{2-\gamma_{i-1}}$; and c_i's are some positive constants that are chosen in such a way that they guarantee the polynomial $p(r) = r^n + c_1 r^{n-1} + c_2 r^{n-2} + \ldots + c_n r^0$ being Hurwitz.

To guarantee finite-time convergence of s (defined in (3.40)) to zero the following surface is considered [9]:

$$\sigma = \dot{s} + \alpha s; \alpha > 0 \tag{3.41}$$

where α is a tuning constant. The nominal controller is calculated considering $\dot{\sigma} = 0$.

$$\begin{aligned}
\dot{\sigma} = 0 = \ddot{s} + \alpha\dot{s} = \ddot{x}_1 + \sum_{i=1}^{n} c_i\gamma_i \left|\tilde{x}^{(n-i)}\right|^{\gamma_i-1} \tilde{x}^{(n-i+1)})\\
+\alpha\left(\dot{\tilde{x}}_1 + \sum_{i=1}^{n} c_i\left|\tilde{x}^{(n-i)}\right|^{\gamma_i} \operatorname{sign}\tilde{x}^{(n-i)}\right) = 0
\end{aligned}$$

$$\hat{u} = -g^{-1}\left[-x_{1d}^{(n)} + f + \left(\frac{1}{1+c_1\gamma_1}\left(\left|\tilde{x}^{(n-1)}\right|^{1-\gamma_1}\right)\right.\right.$$
$$\left.\left.\left[\ddot{x}_1 + \ldots + \alpha\left(\dot{\tilde{x}}_1 + \sum_{i=1}^{n} c_i\left|\tilde{x}^{(n-i)}\right|^{\gamma_i}\operatorname{sign}\tilde{x}^{(n-i)} + \ldots + \sum_{i=2}^{n}\gamma_i c_i\left|\tilde{x}^{(n-i)}\right|^{\gamma_i-1}\tilde{x}^{(n-i+1)}\right)\right]\right)\right] \tag{3.42}$$

The INTSMC for nth-order system is defined as follows:

$$u = \hat{u} - \frac{k}{g}\operatorname{sign}(\sigma) \tag{3.43}$$

Theorem 3.6
The INTSMC, proposed in (3.43), stabilizes system (3.39) for

$$k = \eta + D.$$

Proof: Consider the Lyapunov function $V = \frac{1}{2}\sigma^2$. To prove the stability, the reachability condition should be satisfied:

$$\dot{V} = \frac{d}{dt}\left(\frac{1}{2}\sigma^2(t)\right) = \sigma(t)\dot{\sigma}(t) \le -\eta|\sigma(t)|.$$

$$\dot{V} = \sigma\dot{\sigma} = \sigma\left[c_1\gamma_1\left|\tilde{x}^{(n-1)}\right|^{\gamma_1-1}\left(f + gu + d - x_{1d}^{(n)}\right)\right.$$
$$\left. + \ddot{x}_1 + \alpha\left(\dot{\tilde{x}}_1 + \sum_{i=1}^{n} c_i\left|\tilde{x}^{(n-i)}\right|^{\gamma_i}\operatorname{sign}\tilde{x}^{(n-i)}\right) + \ldots + \sum_{i=2}^{n}\gamma_i c_i\left|\tilde{x}^{(n-i)}\right|^{\gamma_i-1}\tilde{x}^{(n-i+1)}\right]$$

Replacing the controller (3.43) yields:

$$\dot{V} = \sigma\left\{c_1\gamma_1\left|\tilde{x}^{(n-1)}\right|^{\gamma_1-1}(d-(D+\eta)\operatorname{sign}(\sigma))\right\} \le -c_1\gamma_1\left|\tilde{x}^{(n-1)}\right|^{\gamma_1-1}\eta|\sigma|$$

The above relation is always negative and just becomes zero in $\sigma = 0$. So, the controller stabilizes system (3.39).

3.8 MATLAB Simulation

To illustrate and compare the efficiency of proposed controlling methods, in this book the simple and basic structure of an inverted pendulum is considered. This structure is usually used in related literature books.

Considering the inverted pendulum structure, previously described in Section 3.2, as follows:

$$\dot{x}_1 = x_2$$

$$\dot{x}_2 = f(x) + g(x)u + d$$

$$f(x) = \frac{g\ \sin x_1 - mlx_2^2\ \cos x_1\ \sin x_1/(m_c + m)}{l\left(\frac{4}{3} - m\ \cos^2 x_1/(m_c + m)\right)}$$

$$g(x) = \frac{\cos x_1/(m_c + m)}{l\left(\frac{4}{3} - m\ \cos^2 x_1/(m_c + m)\right)} \tag{3.44}$$

The TSMC and its related improvements including FTSMC, NTSMC, FNTSMC, and INTSMC will be designed and discussed in this section.

3.8.1 Simulation Using MATLAB mfiles

To design the terminal sliding surface, Eq. (3.3) is considered and TSMC is calculated by (3.9). q is supposed to be 3 and $p = 5$. The parameters of the inverted pendulum are also considered as $g = \frac{9.8m}{s^2}$, $m_c = 1, m = 0.1$ kg, $l = 0.5$ m.

The equation number for sliding surface and controller structure, related to each controlling method, is also defined in the mfile below.

```
CONTROL = 'NFTSMC';% TSMC,FTSMC,NTSMC,FNTSMC,ITSMC


switch CONTROL
    case 'TSMC'
        METHOD = @TSMCMETHOD;
    case 'FTSMC'
        METHOD = @FTSMCMETHOD;
    case 'NTSMC'
        METHOD = @NTSMCMETHOD;
    case 'NFTSMC'
        METHOD = @NFTSMCMETHOD;
    case 'ITSMC'
        METHOD = @ITSMCMETHOD;
end
```

In the following part different controlling methods are addressed, and accordingly the program switches to the related function.

To execute the controller and plot the results, the following part is utilized.

```
%Plot states, error, controller and sliding surface
 [t,x] = ode23(METHOD,[0 20],[0.5 0 ]);
 xd = sin(pi*t/2);
 xd_d = (pi/2)*(cos(pi*t/2));
```

```
figure(1)
  subplot(211)
  plot( t,x(:,1),':r',t,xd,'k','linewidth',2)
  xlabel('t(s)');ylabel('x1');legend('x1','x1 desired');
  subplot(212)
  plot( t,x(:,2),':r',t,xd_d,'k','linewidth',2)
  xlabel('t(s)');ylabel('x2');legend('x2','x2 desired')
```

Finally the related MATLAB functions for each controlling method are defined as follows:

```
%TSMC controlling method
 function dx = TSMCMETHOD(t,x)
    dx = zeros(2,1);

g = 9.8;mc = 1.0;m = 0.1;l = 0.5;landa;M = 1;L = 0.5;g =
9.8;k = 15;q = 3;p = 5;k = 5;
h = l*(4/3-m*(cos(x(1)))^2/(mc+m));
xd = sin(pi*t/2);
xd_d = (pi/2)*(cos(pi*t/2));xd_dd= -(pi^2/4)*(sin(pi*t/2));
fx = g*sin(x(1))-m*l*(x(2))^2*cos(x(1))*sin(x(1))/(mc+m);
fx = fx/h;
gx = cos(x(1))/(mc+m);gx = gx/h;gxhat = gx; fxhat = fx;
s = x(2)-xd_d +landa*(x(1)-xd)^(q/p); %%equation 3.3
U = -inv(gxhat)*(fxhat+landa*(q/p)*x(2)*(x(1))^((q/p)-1))-
k*sign(s);%%equation 3.9
          dx(1) = x(2);
          dx(2) = fx+gx*U;
 end

%NTSMC controlling method
    function dx = NTSMCMETHOD(t,x)
    dx = zeros(2,1);

g = 9.8;mc = 1.0;m = 0.1;l = 0.5;landa = 1;M = 1;L = 0.5;g
= 9.8;k = 15;q = 3;p = 5;k = 5;
    h = l*(4/3-m*(cos(x(1)))^2/(mc+m));
    xd = sin(pi*t/2); xd_d = (pi/2)*(cos(pi*t/2));xd_dd =
-(pi^2/4)*(sin(pi*t/2));
    e = x(1)-xd;de = x(2)-xd_d;
    T1 = landa*de^(p/q);
    T2 = (de)^(2-(p/q));
    s = e+T1;% equation 3.15
    x1 = x(1);x2 = x(2);
   fx = ((M+m)*g*sin(x1)-m*L*(x2^2)*sin(x1)*cos(x1))/((4/3)
```

```
*(M+m)*L-m*L*(0.5+0.5*cos(2*x1)));
    gx = cos(x1)/((4/3)*(M+m)*L-m*L*(0.5+0.5*cos(2*x1)));
    U = (-1/gx)*(fx-xd_dd+(q/(p*landa))*T2+k*sign(s));%
equation 3.17
     dx(1) = x(2);
     dx(2) = fx+gx*U;
    end

%FTSMC controlling method
    function dx = FTSMCMETHOD(t,x)
    dx = zeros(2,1);

g = 9.8;mc = 1.0;m = 0.1;l = 0.5;alpha = 1;beta = 1;M = 1;L
= 0.5;g = 9.8;k = 15;q = 3;p = 5;k = 5;
    h = l*(4/3-m*(cos(x(1)))^2/(mc+m));

    xd = sin(pi*t/2); xd_d = (pi/2)*(cos(pi*t/2));xd_dd =
-(pi^2/4)*(sin(pi*t/2));
    e = x(1)-xd;de = x(2)-xd_d;
    T1 = beta*e^(q/p);
    T2 = (e)^(q/p-1);
    s = de+alpha*e+T1;% equation 3.19
    x1 = x(1);x2 = x(2);
   fx = ((M+m)*g*sin(x1)-m*L*(x2^2)*sin(x1)*cos(x1))/((4/3)
*(M+m)*L-m*L*(0.5+0.5*cos(2*x1)));
    gx = cos(x1)/((4/3)*(M+m)*L-m*L*(0.5+0.5*cos(2*x1)));
    U = (-1/gx)*(fx-xd_dd+alpha*de+beta*q/p*T2*de+k*sign
(s));%equation 3.25
     dx(1) = x(2);
     dx(2) = fx+gx*U;
    end

%FNTSMC controlling method
    function [U,dx] = NFTSMCMETHOD(t,x)
    dx = zeros(2,1);
   g = 9.8;mc = 1.0;m = 0.1;l = 0.5;alpha = 1;beta = 1;M =
1;L = 0.5;g = 9.8;k = 15;q = 3;p = 5;k = 5;
h = l*(4/3-m*(cos(x(1)))^2/(mc+m));
xd=sin(pi*t/2);xd_d=(pi/2)*(cos(pi*t/2));xd_dd=-(pi^2/4)
*(sin(pi*t/2));
    e = x(1)-xd;de = x(2)-xd_d;
    T1 = (1/beta)*de^(p/q);
    T2 = (1+(1/alpha))*(de)^(2-(p/q));
    s = e+(1/alpha)*e+T1;% equation 3.31
```

```
    x1 = x(1);x2 = x(2);
fx = ((M+m)*g*sin(x1)-m*L*(x2^2)*sin(x1)*cos(x1))/((4/3)*(M
+m)*L-m*L*(0.5+0.5*cos(2*x1)));
    gx = cos(x1)/((4/3)*(M+m)*L-m*L*(0.5+0.5*cos(2*x1)));
    U = (-1/gx)*(fx-xd_dd+T2/(p/beta*q)+k*sign(s));%
equation 3.32
     dx(1) = x(2);
     dx(2) = fx+gx*U;
    end
```

Figures 3.2 and 3.3 show the results of applying TSMC on inverted pendulum.

Comparing Figures 3.3 and 2.12 shows that TSMC converges faster and with less chattering in comparison with classic SMC. Systems states converge to their desired value in about 1.9 seconds.

Figures 3.4 and 3.5 show the NTSMC results.

Comparing controlling signal in Figures 3.3 and 3.5 shows that although the tracking error in NTSMC is more than TSMC, it effectively eliminates the sudden rises in controlling value [10].

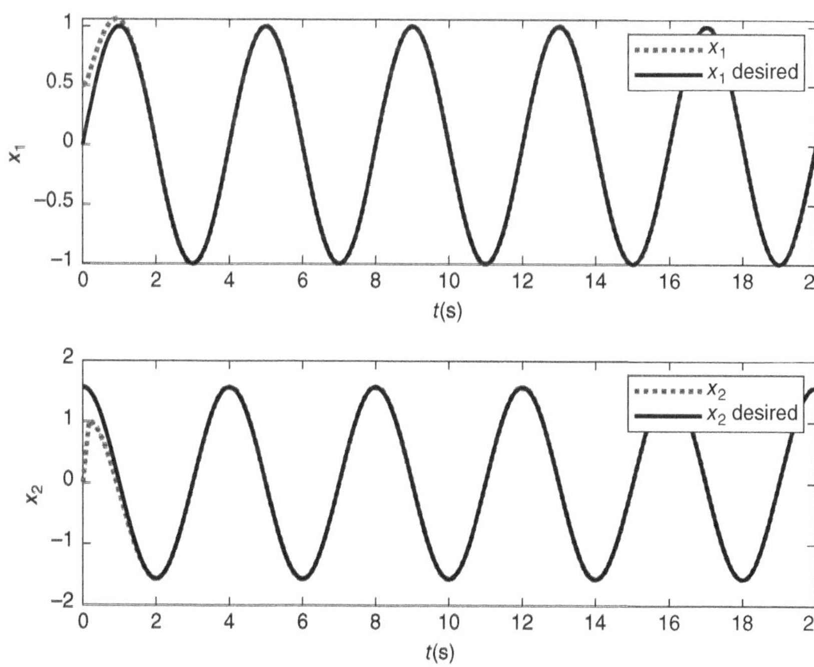

Figure 3.2 Inverted pendulum position tracking using TSMC.

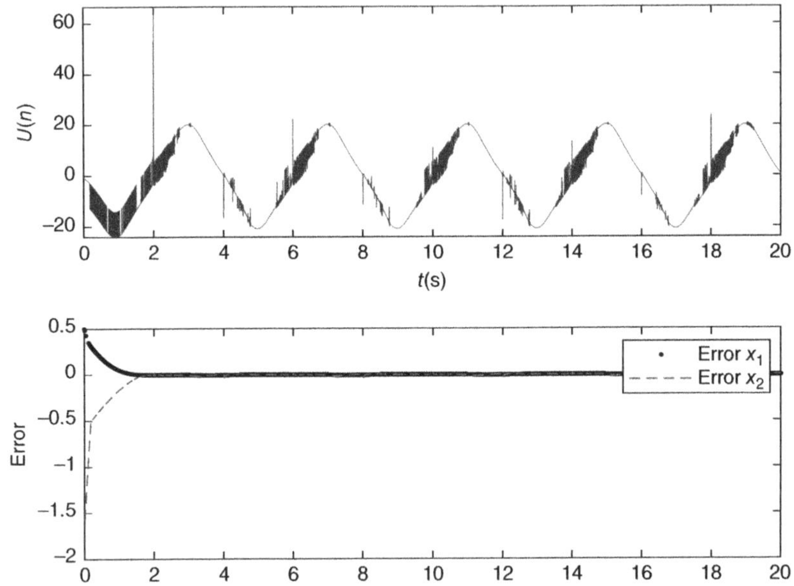

Figure 3.3 Inverted pendulum controller and tracking error using TSMC.

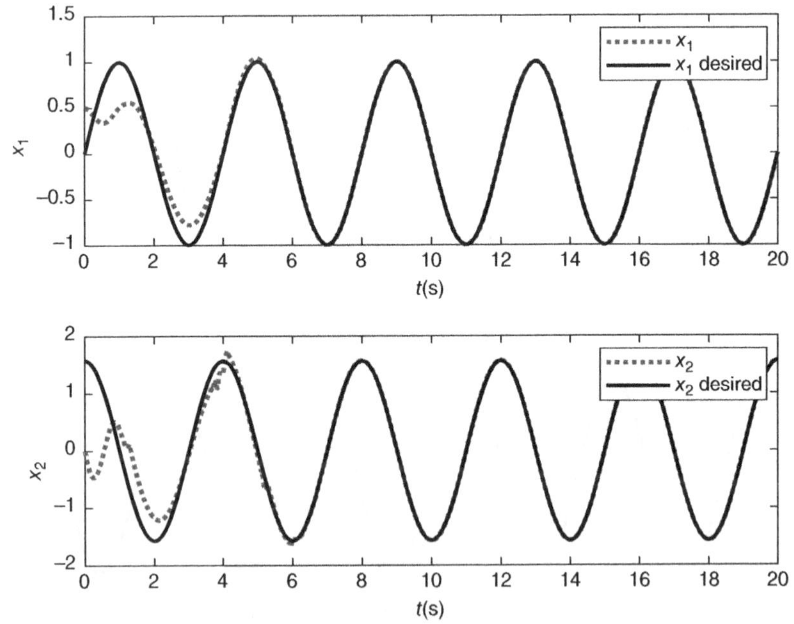

Figure 3.4 Inverted pendulum position tracking using NTSMC.

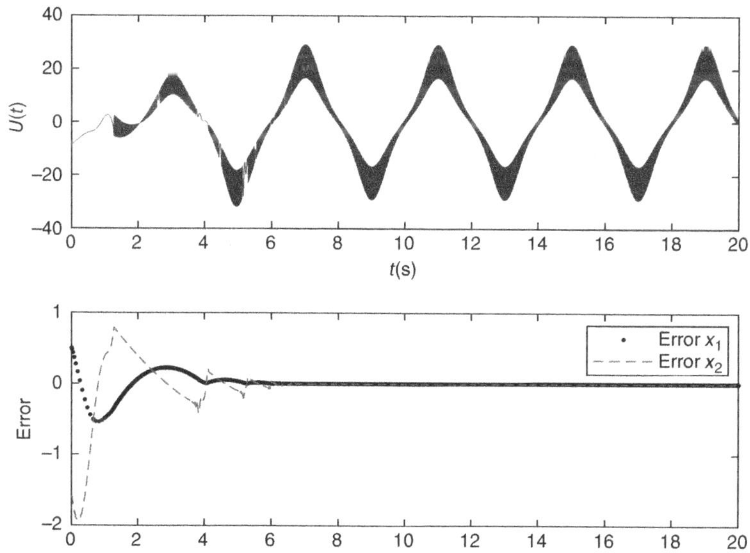

Figure 3.5 Inverted pendulum controller and tracking error using NTSMC.

Figures 3.6 and 3.7 illustrate the tracking efficiency using FTSMC.

As it is illustrated, FTSMC effectively reduces the convergence time to desired signal.

Finally the results of NFTSMC are proposed in Figures 3.8 and 3.9, which not only contains the advantage of fast convergence of FTSMC but also avoids the singularity problem.

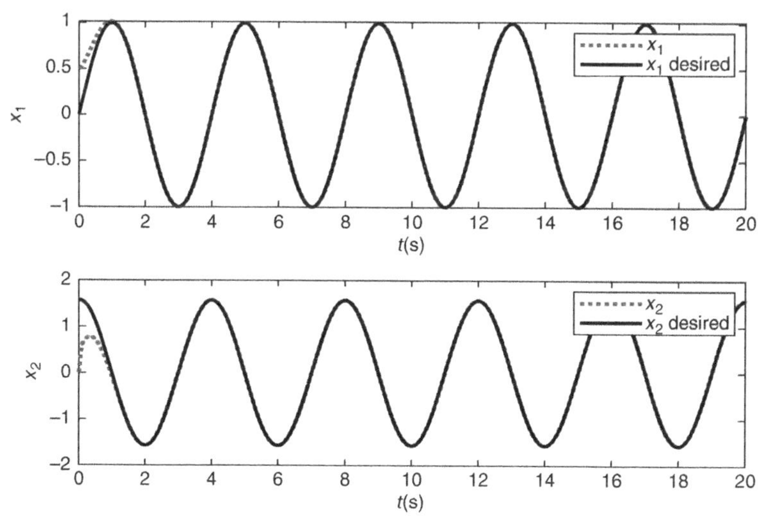

Figure 3.6 Inverted pendulum position tracking using FTSMC.

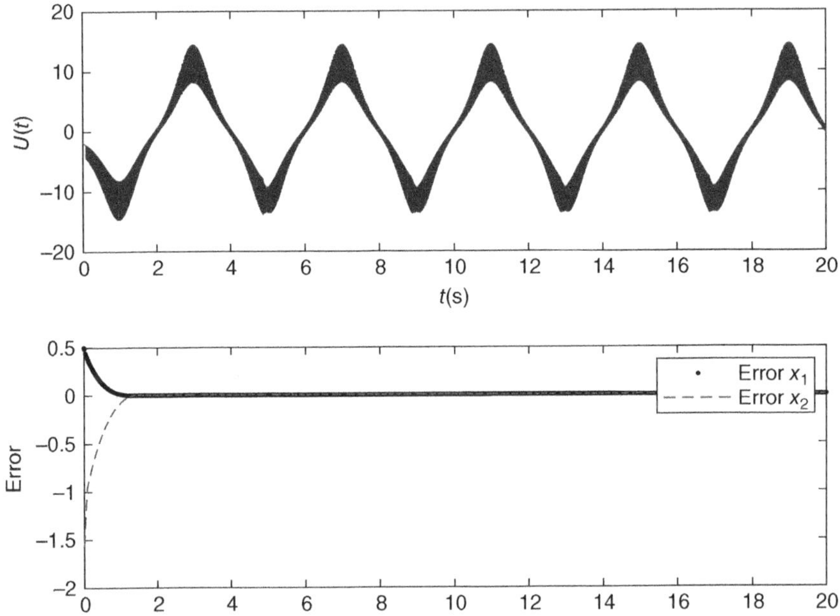

Figure 3.7 Inverted pendulum controller and tracking error using FTSMC.

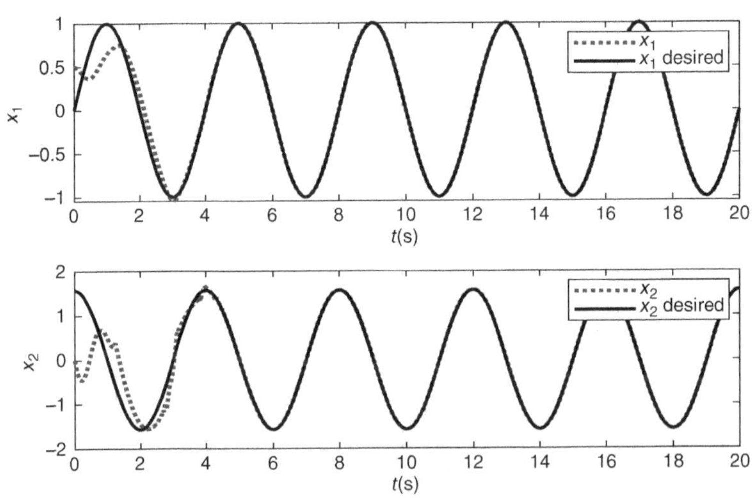

Figure 3.8 Inverted pendulum position tracking using NFTSMC.

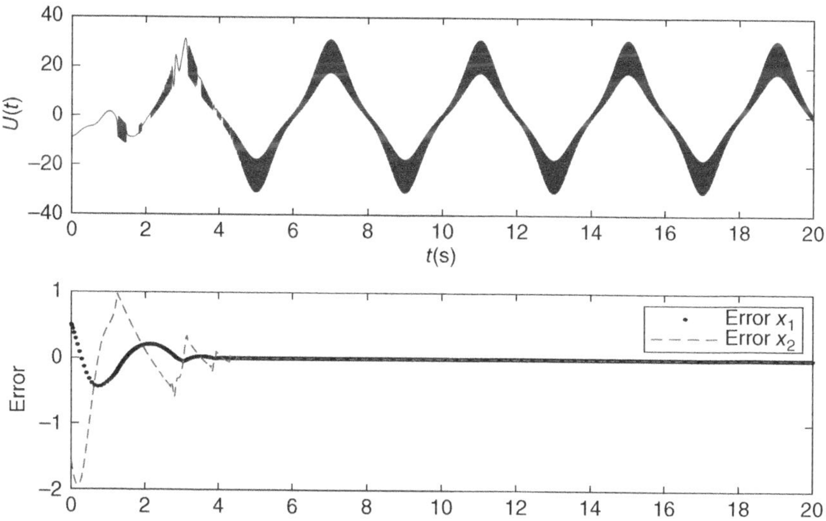

Figure 3.9 Inverted pendulum controller and tracking error using NFTSMC.

3.8.2 Designing Improved Sliding Mode Controllers for Inverted Pendulum in SIMULINK

To design TSMC, NTSMC, FTSMC, NFTSMC, and INTSMC we consider the remarks proposed in this chapter. The controlling block diagram is defined in Figure 3.10. As it is shown it contains a controller block which includes sliding surface and controlling signal and varies from one method to another one.

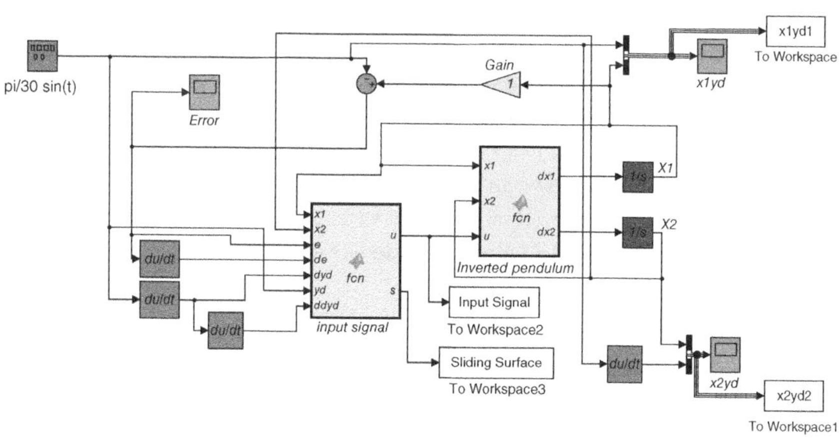

Figure 3.10 Block diagram of improved terminal sliding mode control of inverted pendulum.

The second block defines the inverted pendulum dynamic, and it is fixed.

The inverted pendulum function is defined as follows. It should be mentioned that the inverted pendulum dynamic function, defined below, and Simulink model, proposed in Figure 3.10, are repeated for all controlling methods including TSMC, NTSMC, FTSMC, and NFTSMC. The block diagram related to INTSMC is a little different as shown in Figure 3.15.

```
function [dx1,dx2] = fcn(x1,x2,u)

%THIS FUNCTION DEFINES THE INVERTED PENDULUM BLOCK IN
FIGURE 3.10
m = 0.1; M = 1; L = 0.5; g = 9.8;
zeros(2,1);
dx1 = x2;
f = ((M+m)*g*sin(x1)-m*L*(x2^2)*sin(x1)*cos(x1))/((4/3)*
    (M+m)*L-m*L*(0.5+0.5*cos(2*x1)));
g =  cos(x1)/((4/3)*(M+m)*L-m*L*(0.5+0.5*cos(2*x1)));
dx2 = f+b*u;
```

The values of the parameters used in the simulations are as follows:

$$k = 25, \lambda = 5, p = 5, q = 3, \alpha = 1, \beta = 1, x_0 = [0.2{,}0.1], x_d = \frac{\pi}{30}\sin(t)$$

The following function defines the input signal MATLAB function which contains TSMC controller as defined in (3.9) and Remark 3.1 (Figure 3.11).

```
function [u,s] = fcn(x1,x2,e,de,dyd,yd,ddyd)

%THIS FUNCTION ILLUSTRATES THE TSMC DETAILS

 m = 0.1;M = 1;L = 0.5;g = 9.8;k = 15;landa = 5;p = 5;q =
 T1 = abs(e)^(q/p)*sign(de);
 T2 = abs(e)^(q/p -1)*sign(e);
 s = de+landa*T1;
 f = ((M+m)*g*sin(x1)-m*L*(x2^2)*sin(x1)*cos(x1))/((4/3)*
(M+m)*L-m*L*(0.5+0.5*cos(2*x1)));
 g = cos(x1)/((4/3)*(M+m)*L-m*L*(0.5+0.5*cos(2*x1)));
 u = (-1/g)*(f+(q/p)/landa*T2*de+k*sign(s));
 end
```

As it is shown, the states converge to desired value in 1.9 seconds, which is similar to Figure 3.3. Using Remark 3.1 results in more evident chattering phenomena.

In MATLAB function below, the NTSMC signal is proposed using Eq. (3.17) and Remark 3.2.

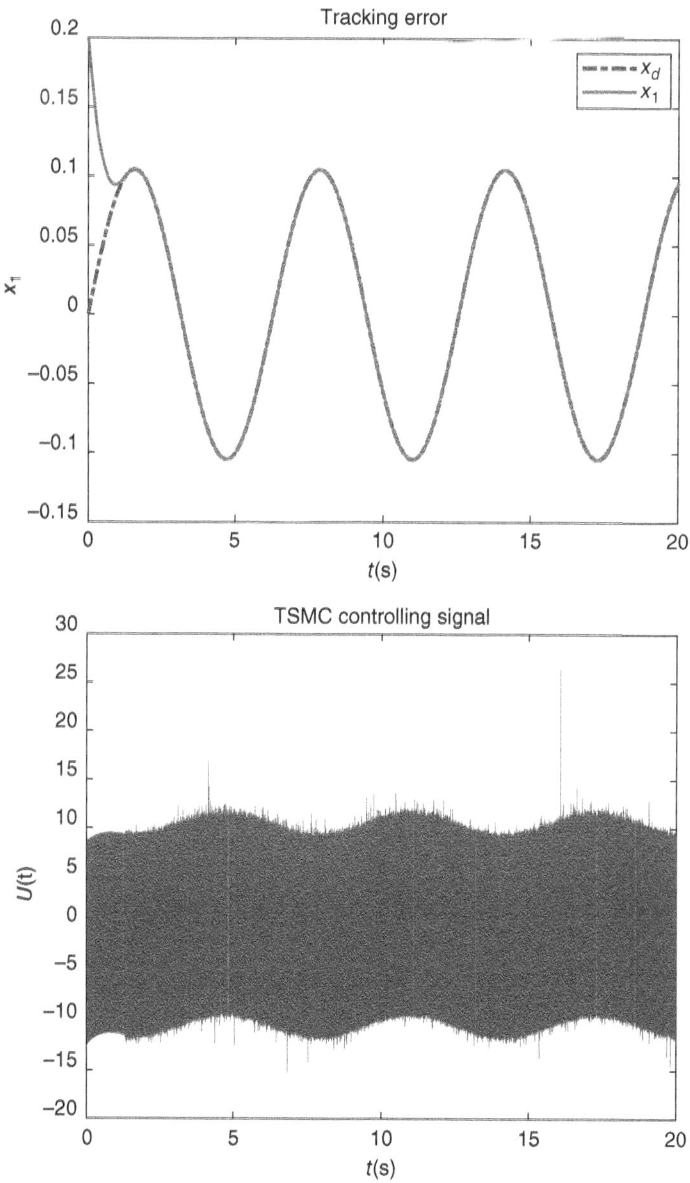

Figure 3.11 Tracking error and TSMC controlling signal.

```
function [u,s] = fcn(x1,x2,e,de,dyd,yd,ddyd)

%THIS FUNCTION ILLUSTRATES THE NTSMC DETAILS

 m = 0.1;M = 1;L = 0.5;g = 9.8;k = 15;landa = 5;p = 5;q =
 T1 = abs(de)^(p/q)*sign(de);
 T2 = abs(x2)^(2-p/q)*sign(x2);
 s = e+landa*T1;
 f = ((M+m)*g*sin(x1)-m*L*(x2^2)*sin(x1)*cos(x1))/((4/3)*
(M+m)*L-m*L*(0.5+0.5*cos(2*x1)));
 g = cos(x1)/((4/3)*(M+m)*L-m*L*(0.5+0.5*cos(2*x1)));
 u = (-1/b)*(f+(q/p)/landa*T2+k*sign(s));
 end
```

Figure 3.12 shows the result of applying NSTSMC on inverted pendulum.

As it is shown, although the peak-to-peak controller variations reduce to 20 N, the convergence time increases to about 4 seconds, compared with TSMC.

The following MATLAB function defines the input signal block for FTSMC using Remark 3.3.

```
function [u,s] = fcn(x1,x2,e,de,dyd,yd,ddyd)

m = 0.1;M = 1;L = 0.5;g = 9.8;k = 15;beta = 1;p = 5;q = 3,
alpha = 1;

T1 = beta*abs(e)^(q/p)*sign(e);
T2 = abs(x1)^(q/p-1)*sign(x1);

s = de+alpha*e+T1;

f = ((M+m)*g*sin(x1)-m*L*(x2^2)*sin(x1)*cos(x1))/((4/3)*
(M+m)*L-m*L*(0.5+0.5*cos(2*x1)));

b = cos(x1)/((4/3)*(M+m)*L-m*L*(0.5+0.5*cos(2*x1)));

u = (-1/b)*(f+alpha*x2+beta*q/p*T2*x2+k*sign(s));
end
```

Figure 3.13 shows the result of applying FTSMC on inverted pendulum.

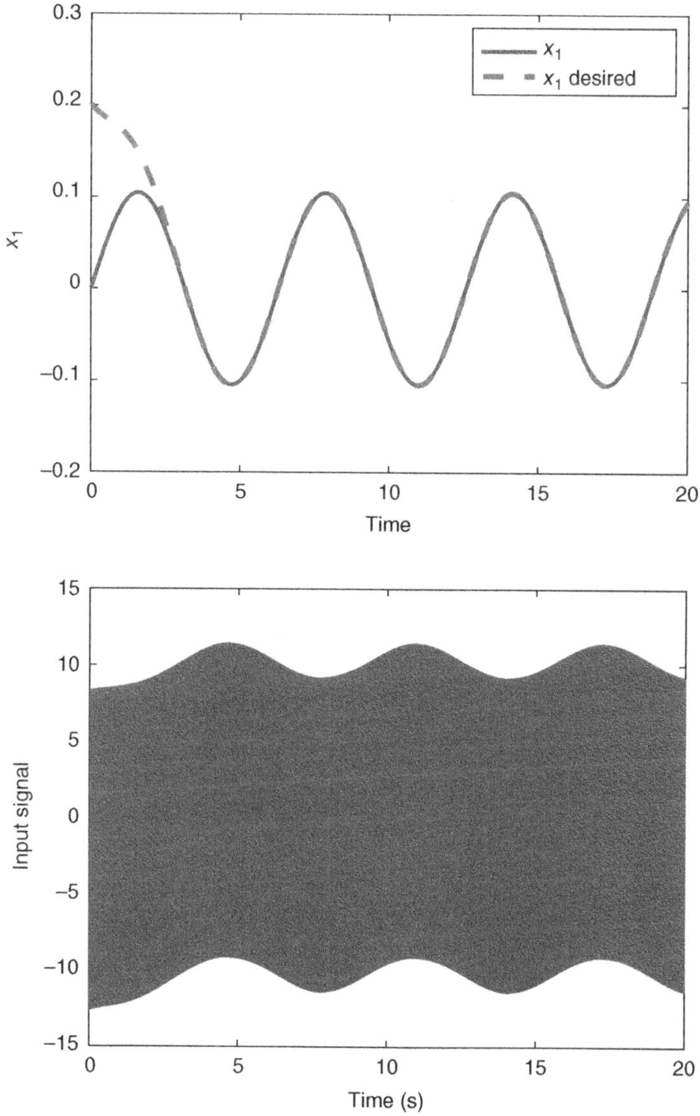

Figure 3.12 Inverted pendulum position tracking and controller using NTSMC.

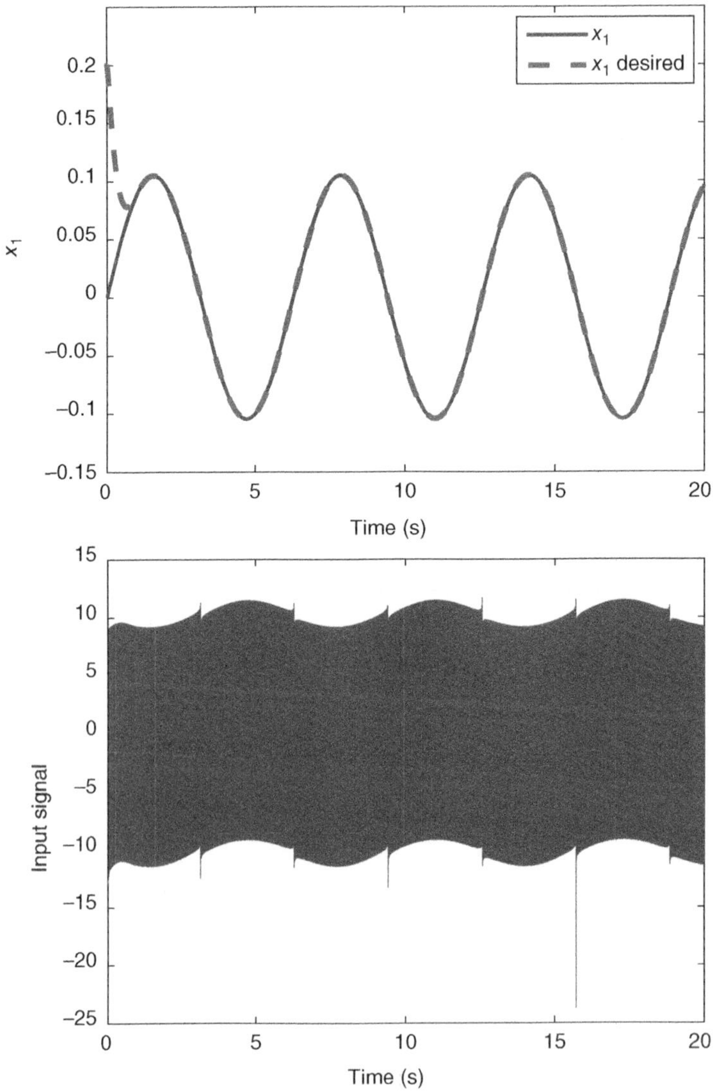

Figure 3.13 Inverted pendulum position tracking and FTSMC signal.

As it is shown, FTSMC reduces the convergence time effectively. But the increase of chattering phenomena is not avoidable. To improve the results an NFTSMC is proposed in Figure 3.14:

```
function [u,s] = fcn(x1,x2,e,de,dyd,yd,ddyd)
m = 0.1;M = 1;L = 0.5;g = 9.8;k = 15;beta = 1;p = 5;q = 3,
alpha = 1;

T1 = 1/beta*abs(de)^(p/q)*sign(de);
T2 = abs(x2)^(2-p/q)*sign(x2);

s = e+1/alpha*e+T1;
f = ((M+m)*g*sin(x1)-m*L*(x2^2)*sin(x1)*cos(x1))/((4/3)*
(M+m)*L-m*L*(0.5+0.5*cos(2*x1)));

b = cos(x1)/((4/3)*(M+m)*L-m*L*(0.5+0.5*cos(2*x1)));

u = (-1/b)*(f+((1+1/alpha)/1/beta*p/q)*T2+k*sign(s));
end
```

Figure 3.15 shows the block diagram of INTSMC. As it is shown, it contains three MATLAB functions. As the inverted pendulum dynamic is the same as previous part, only the two remaining functions will be presented in the sequel.

The input signal MATLAB function is as follows. It is clear that parameters such as $c_1$and c_2 should be adjusted precisely to confirm the proper performance of system. Sliding surface function contains the part which should be integrated, as defined in (3.35), and is proposed as follows:

```
function s_initial = fcn(e,de)

gama1 = 1.5;gama2 = gama1/(2-gama1);c1 = 4.6;c2 = 20;

T2 = c1*(abs(de)^(gama1))*sign(de);
T3 = c2*(abs(e)^(gama2))*sign(e);

 s_initial = T2+T3;
end
```

And the INTSMC MATLAB function is defined, considering relation (3.38), as follows:

```
function [u,s] = fcn(x1,x2,e,de,s_in,ddyd)

m = 0.1;M = 1;L = 0.5;g = 9.8;k = 15;alpha = 1;
gama1 = 1.5;gama2 = gama1/(2-gama1);c1 = 4.6;c2 = 20;
```

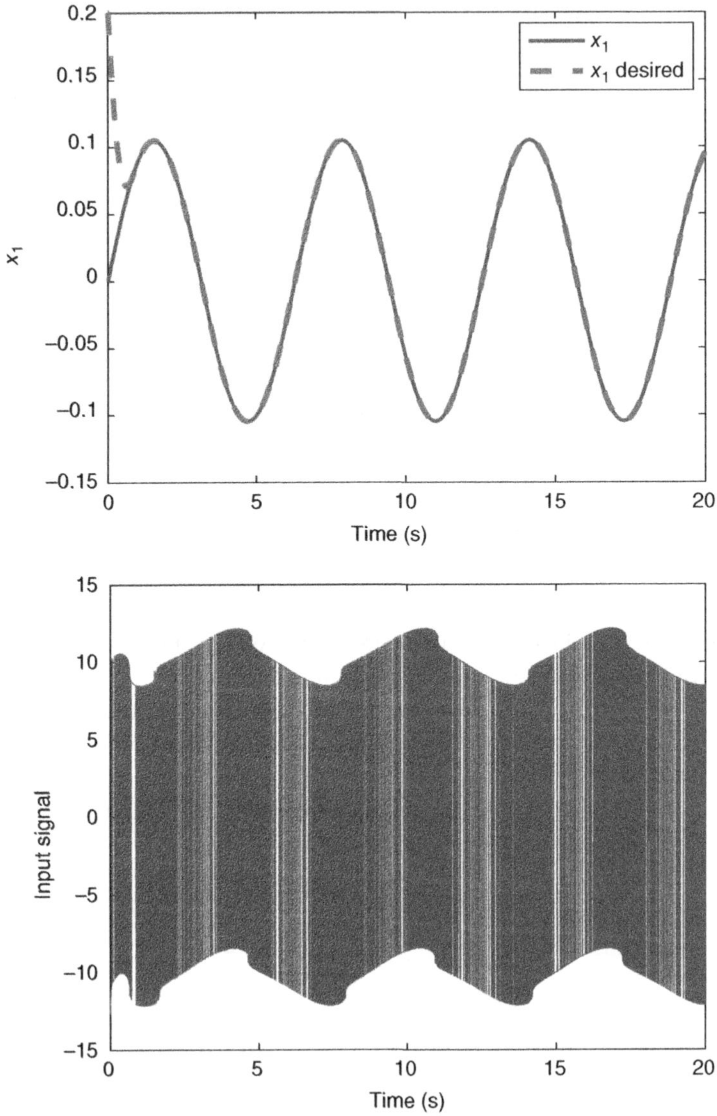

Figure 3.14 Inverted pendulum position tracking using NFTSMC and its related controlling signal.

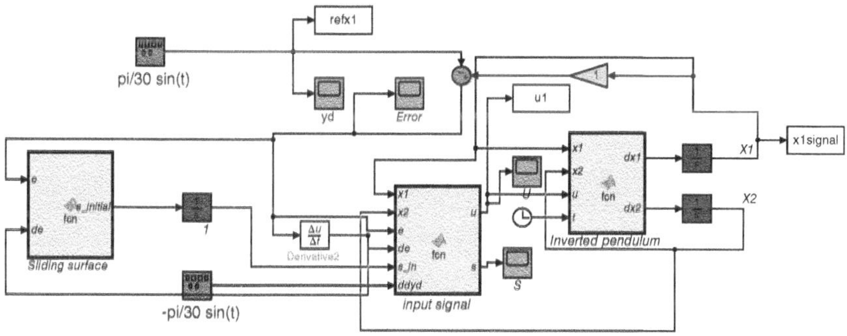

Figure 3.15 INTSMC block diagram.

```
T1 = 1+c1*gama1*abs(de)^(gama1-1);
T2 = c1*(abs(de)^(gama1))*sign(de);
T3 = c2*(abs(e)^(gama2))*sign(e);
s = e+s_in;
sig = de+T2+T3+alpha*s;
f = ((M+m)*g*sin(x1)-m*L*(x2^2)*sin(x1)*cos(x1))/((4/3)*
(M+m)*L-m*L*(0.5+0.5*cos(2*x1)));

b = cos(x1)/((4/3)*(M+m)*L-m*L*(0.5+0.5*cos(2*x1)));

u = (-1/b)*(f-ddyd+(alpha/T1)*(de+T2+T3)+(1/T1)*c2*gama2*
(de)*abs(e)^(gama2-1))-(k/b)*sign(sig);

end
```

The simulation results of INTSMC implementation are illustrated in Figure 3.16.

3.9 Conclusion

Considering simulation results it is concluded that using nonlinear sliding surface in terminal sliding mode control and fast TSMC effectively reduces the convergence time and, to reduce the chattering phenomena, INTSMC can be applied. In addition, using NTSMC also decreases the sudden unwanted jumps in controlling signal. The following table clarifies the concluded results.

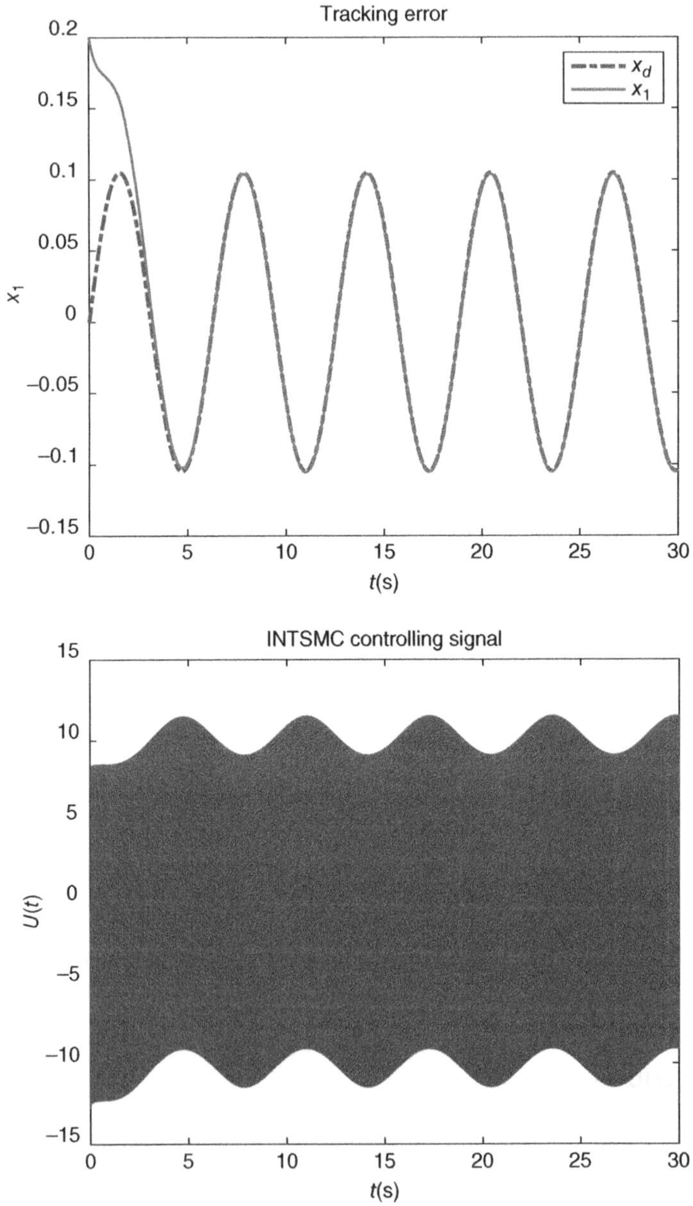

Figure 3.16 Inverted pendulum position tracking and INTSMC signal.

	Convergence time(s)	Integral of squared error	Controller peak-to peak max value (*N*)
TSMC	1.275	0.0138	32
NTSMC	1.18	0.0131	20.5
FTSMC	0.95	0.0092	20.7
NFTSMC	0.9	0.0076	20.5
INTSMC	4.2	0.336	20.4

References

1. Hou, Z., Wang, Y, and Liu, L. (2014). Improved nonsingular fast terminal sliding mode controller for nonlinear systems. *2014 International Conference on Mechatronics and Control (ICMC)* (3–5 July 2014), Jinzhou, China. https://doi.org/10.1109/ICMC.2014.7231640
2. Xinghuo, Y. (2020). Terminal sliding mode control – an overview. *IEEE Open Journal of the Industrial Electronics Society* 2 (1): 36–52. https://doi.org/10.1109/OJIES.2020.3040412.
3. Moghaddam, R.K. and Mohamadie, N.K. (2023). Terminal nonsingular sliding mode control of wave energy converters for maximum power absorption. *Ocean Engineering* 281 (1): 114473.
4. Kardehi Moghaddam, R. and Baratpoor, J. (2023). Fuzzy adaptive nonsingular terminal sliding mode control of a miniature helicopter. *Journal of Aerospace Information Systems* 1–12. https://doi.org/10.2514/1.I011156.
5. Rabbani, M. and Kardehi Moghaddam, R. (2024). A newly proposed super-twisting backstepping. *International Journal of Nonlinear Analysis and Applications* 4: 295–309.
6. Xu, B., Zhang, L., and Ji, W. (2021). Improved non-singular fast terminal sliding mode control with disturbance observer for PMSM drives. *IEEE Transactions on Transportation* 7 (4).
7. Fenga, Y., Yub, X., and Manc, Z. (2002). Non-singular terminal sliding mode control of rigid manipulators. *Automatica* 38: 2159–2167. https://doi.org/10.1016/S0005-1098(02)00147-4.
8. Shen, W. and Wang, J. (2022). An integral terminal sliding mode control scheme for speed control system using a double-variable hydraulic transformer. *ISA Transactions* 124: 386–394. https://doi.org/10.1016/j.isatra.2019.08.068.

9 Yiguang, H., Yang, G., Cheng, D., and Spurgeon, S. (2004). Finite time convergent control using terminal sliding mode. *Journal of Control Theory and Applications* 2: 69–74. https://doi.org/10.1007/s11768-004-0026-6.

10 Xu, B., Zhang, L., and Ji, W. (2021). Improved non-singular fast terminal sliding mode control with disturbance observer for PMSM drives. *IEEE Transactions on Transportation Electrification* 7 (4). https://doi.org/10.1109/TTE.2021.3083925

11 Mohammadi, R., Hagh, Y.S., Palm, R.H., and Handroos, H. Integral non-singular terminal sliding mode controller for nth-order nonlinear systems. *IEEE Access* 7: 102792–102802. https://doi.org/10.1109/ACCESS.2019.2930798.

4

Dynamic Sliding Mode Control

4.1 Introduction

In the design of a sliding mode controller for a system, it is sometimes necessary to incorporate additional dynamics into the sliding surface structure. This approach aims to enhance the stability of the control system and sliding mode, ultimately achieving the desired response and behavior for the system [1]. In the dynamic sliding mode control (DSMC) methods, it is assumed that all the system states or equivalently, the output derivatives with appropriate degrees, are known in order to design the control law [2].

In this chapter, a dynamic sliding surface is used to design the sliding control law. First, a classic sliding surface is defined based on the tracking error and its derivatives, and then, using it, a dynamic sliding surface is proposed. The sliding surface structure is first order or higher order depending on the derivatives in the control input. Also, discontinuities can be transferred to the first or higher-order derivative at the control input. As a result, a DSMC is obtained, which significantly reduces the chattering phenomenon [3].

4.2 Designing the Dynamic Sliding Mode Controller

4.2.1 Statement of the Problem

Consider the following nonlinear system:

$$\begin{cases} \dot{x}_1 = x_2 \\ \dot{x}_2 = f(x) + g(x)u + d(t) \\ y = x_1 \end{cases} \tag{4.1}$$

Methods of Developing Sliding Mode Controllers: Design and MATLAB Simulation, First Edition.
Reihaneh Kardehi Moghaddam and Mostafa Rabbani.

Published 2025 by John Wiley & Sons, Inc.

where $x = [x_1, x_2]^T$ is the state vector of the system, $f(x)$ and $g(x) \neq 0$ are nonlinear, known, and smooth functions, u is the control input, and $d(t)$ represents uncertainties, where $|d(t)| \leq D_1$ and $|\dot{d}(t)| \leq D_2$.

The goal of designing a dynamic sliding mode controller is asymptotic convergence of tracking error to zero.

4.2.2 Controller Design and Stability Analysis

For this purpose, if the tracking error is considered as $e = y - y_d$, where y_d is the desired output, then the typical sliding surface is defined as a linear combination of the error and its derivative as follows:

$$s = \dot{e} + k_1 e \tag{4.2}$$

where $k_1 > 0$. In the following, a new dynamic sliding surface is defined as follows:

$$\sigma = \dot{s} + k_2 s \tag{4.3}$$

where $k_2 > 0$.

Theorem 4.1
If the control signal is chosen according to Eq. (4.4), the system's trajectories will reach zero asymptotically [3].

$$\begin{aligned} \dot{u} = g(x)^{-1}\Big(&-\dot{f}(x) + (k_1 + k_2)\ddot{y}_d + \dddot{y}_d - (\dot{g}(x) \\ &+ (k_1 + k_2)g(x))u - (k_1 + k_2)f(x) - k_1 k_2 \dot{e} - K\,\mathrm{sgn}(\sigma)\Big) \end{aligned} \tag{4.4}$$

Proof: Consider the Lyapunov function $V = \frac{1}{2}\sigma^2$; taking the derivative of Lyapunov function results in:

$$\dot{V} = \sigma\dot{\sigma} = \sigma\left(\dot{f}(x) + \dot{g}(x)u + g(x)\dot{u} + \dot{d}(t) - \dddot{y}_d + k_1\ddot{e} + k_2\dot{s}\right) \tag{4.5}$$

Substituting the values of $\ddot{e}$ and $\dot{s}$, (4.5) is rewritten as follows:

$$\begin{aligned} \dot{V} = \sigma\Big(&\dot{f}(x) - (k_1 + k_2)\ddot{y}_d - \dddot{y}_d + \dot{d}(t) + (k_1 + k_2)d(t) + (\dot{g}(x) \\ &+ (k_1 + k_2)g(x))u + (k_1 + k_2)f(x) + \dot{g}(x)u + k_1 k_2 \dot{e}\Big) \end{aligned} \tag{4.6}$$

Inserting (4.4) into (4.6) gives:

$$\dot{V} = \sigma\left(\dot{d}(t) + (k_1 + k_2)d(t) - K\,\mathrm{sgn}(\sigma)\right) \tag{4.7}$$

If K is chosen in such a way that $K > (k_1 + k_2)D_1 + D_2$, it yields:

$$\dot{V} = \sigma\big(\dot{d}(t) + (k_1 + k_2)d(t)\big) - K|\sigma| \leq ((k_1 + k_2)D_1 + D_2)\sigma - K|\sigma| < 0 \tag{4.8}$$

Therefore, the sliding surface reaches zero asymptotically and the proof of the theorem is complete.

Remark 4.1
As can be seen, there is a discontinuity caused by the sign function in the signal $\dot{u}$ and the control effort u will be without chattering.

4.3 DSMC MATLAB Programming

Consider the following system:

$$\begin{cases} \dot{x}_1 = x_2 \\ \dot{x}_2 = -20x^2 + 100u + 2\sin(t) \end{cases} \tag{4.9}$$

If the desired trajectory $y_d = \sin(t)$ and the initial value of the state vector $x(0) = [0.5, 0]^T$ are considered, then by choosing $k_1 = 10$, $k_2 = 20$, $K > (k_1 + k_2) D_1 + D_2 = 70$, the simulation results are obtained as follows.

As seen in Figures 4.1 and 4.2 the path tracking is done well using the proposed controller. On the other hand, it can be seen in Figures 4.3 and 4.4 that although the dynamic controller $\dot{u}$ suffers from chattering, in the real control signal u there is no chattering. The above controller is simulated using the following MATLAB program.

The systems and controller dynamics are defined in the following mfile:

```
function dx = dynamic_sliding(t,x)
dx = zeros(3,1);
% U is considered to be X3
% Desired trajectory and its derivatives
  xd = sin(t);
  xd_d = cos(t);
  xd_dd = -(sin(t));
  k1 = 10;
  k2 = 20;
  k = 70;
%Sliding surface structure equation 4.2
   s = x(2)-xd_d +k1*(x(1)-xd);
```

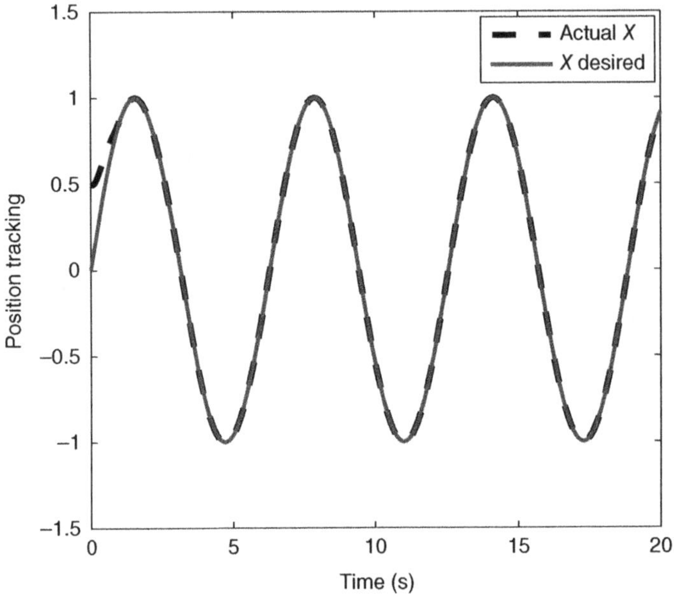

Figure 4.1 Tracking the desired trajectory.

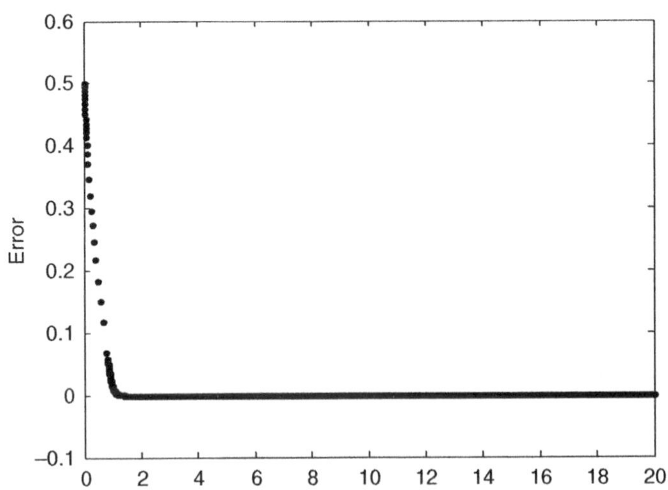

Figure 4.2 Tracking error.

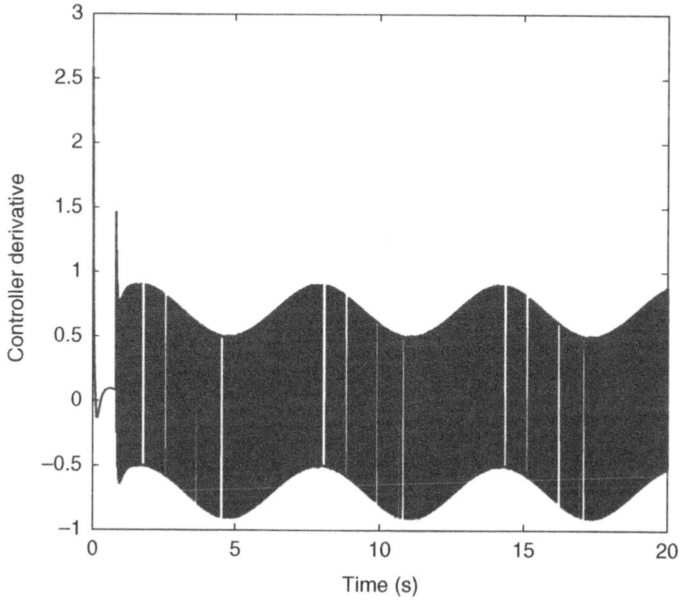

Figure 4.3 Dynamic controller.

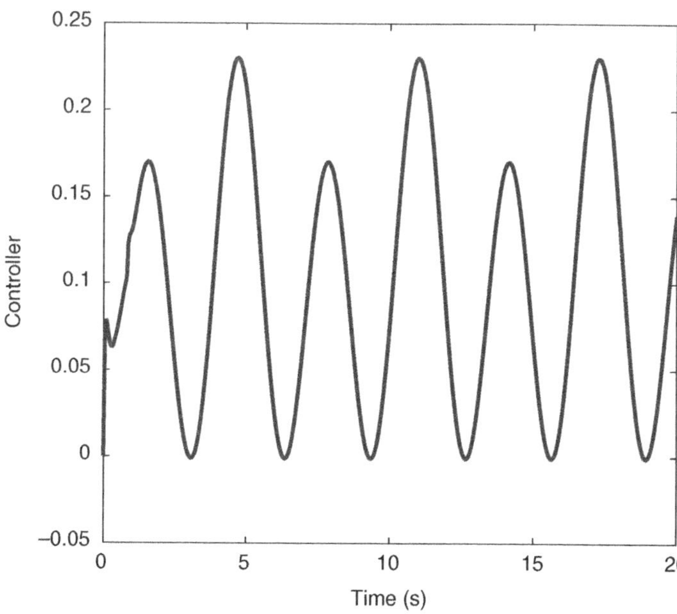

Figure 4.4 Controller $\dot{u}$ that is applied to the system.

```
%SYSTEM'S DYNAMIC
   dx(1) = x(2);
    dx(2) = -20*(x(1)^2)+100*x(3)+2*sin(t);

%equation 4.3 defines sima
  sigma = (-20*(x(1)^2)+100*x(3)+2*sin(t))-
  xd_dd+k1*(x(2)-xd_d)+k2*s;
%Equation 4.4 defines controller
  dx(3) = (1/100)*(-40*x(1)+(k1+k2)*xd_dd-cos(t)-
  (0+(k1+k2)*100)*x(3)-(k1+k2)*( -20*(x(1)^2))-
  k1*k2*(x(2)-xd_d)-k*sign(sigma));
```

To plot the results, the execution mfile below is implemented:

```
%PLOT STATES AND TRACKING ERROR
  [t,x] = ode23(@dynamic_sliding,[0 20],[0.5 0 0]);
  figure(1)
  plot( t,x(:,1),'b--',t,sin(t),'r','linewidth',2)
  xlabel('t(s)')
  ylabel('POSITION TRACKING');
  legend('actual x','desired x')
  x1 = x(:,1);x2 = x(:,2);x3 = x(:,3);u = x3;
figure(2)
plot( t,u,'b')
xlabel('t(s)')
ylabel('controller')
figure(3)
error = x(:,1)- sin(t);
xlabel('t(s)')
plot( t,error,'k.')
ylabel('error')
    k1 = 10;
    k2 = 20;
    k = 70;
[m,n] = size(t);s = zeros(m,1);difu = zeros(m,1);sigma =
zeros(m,1);
for i = 1:m
    xd = sin(t(i));
    xd_d = cos(t(i));
    xd_dd = -(sin(t(i)));
```

```
    s(i) = x2(i)-xd_d +k1*(x1(i)-xd);
    sigma(i) = (-20*(x1(i)^2)+100*x3(i)+2*sin(t(i)))-
    xd_dd+k1*(x2(i)-xd_d)+k2*s(i);
    difu(i) = (1/100)*(-40*x1(i)+(k1+k2)*xd_dd-
    cos(t(i))-(0+(k1+k2)*100)*x3(i)-(k1+k2)*( -
    20*(x1(i)^2))-k1*k2*(x2(i)-xd_d)-k*sign(sigma(i)));
end
figure(4)
plot( t,difu,'b')
xlabel('t(s)')
ylabel(' controller derivative')
```

4.3.1 Programming in Simulink Environment

The system of the previous section is simulated as follows in MATLAB Simulink environment; as you can see, the system model block represents the dynamics of the system, and the controller is defined as the x_3 state, and the sliding surface block represents the sliding surfaces. The mentioned blocks (Figure 4.5) are written by the following functions.

The MATLAB function related to each block is defined as follows:

```
function [s,sigma] = slidingsurface(t,x1,x2,x3)
xd = sin(t);
```

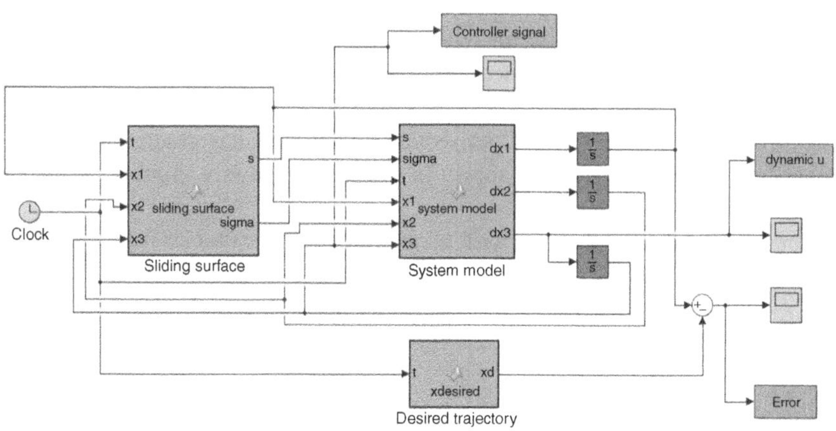

Figure 4.5 Dynamic SMC block diagram.

```
xd_d = cos(t);
xd_dd = -(sin(t));
k1 = 10;
k2 = 20;
k = 70;
s = x2-xd_d +k1*(x1-xd);
sigma = (-20*(x1^2)+100*x3+2*sin(t))-xd_dd+k1*(x2-xd_d)
+k2*s;


function xd = xdesired(t)
xd = sin(t);


function [dx1,dx2,dx3] = systemmodel(s,sigma,t,x1,x2,x3)
xd = sin(t);
xd_d = cos(t);
xd_dd = -(sin(t));
k1 = 10;
k2 = 20;
k = 70;
dx1 = x2;
dx2 = -20*(x1^2)+100*x3+2*sin(t);
dx3 = (1/100)*(-40*x1+(k1+k2)*xd_dd-cos(t)-(0+(k1+k2)
*100)*x3-(k1+k2)*( -20*(x1^2))-k1*k2*(x2-xd_d)-k*sign
(sigma));
```

The dynamic controller and the controller applied to the system described in Simulink with the title (dynamicu) are shown in Figure 4.6, as described in the previous section (Figures 4.3 and 4.4), although the controller does not have high-frequency oscillation, but the dynamic controller has chattering. On the other hand, the tracking error shown in Figure 4.7 shows a rapid convergence to zero.

The following commands are used to call and plot the variables from the Simulink space:

```
[m,n] = size(dynamicu);
t = (0:20/(m-1):20);
plot(t',dynamicu);
plot(t',error);
plot(t', controller);
```

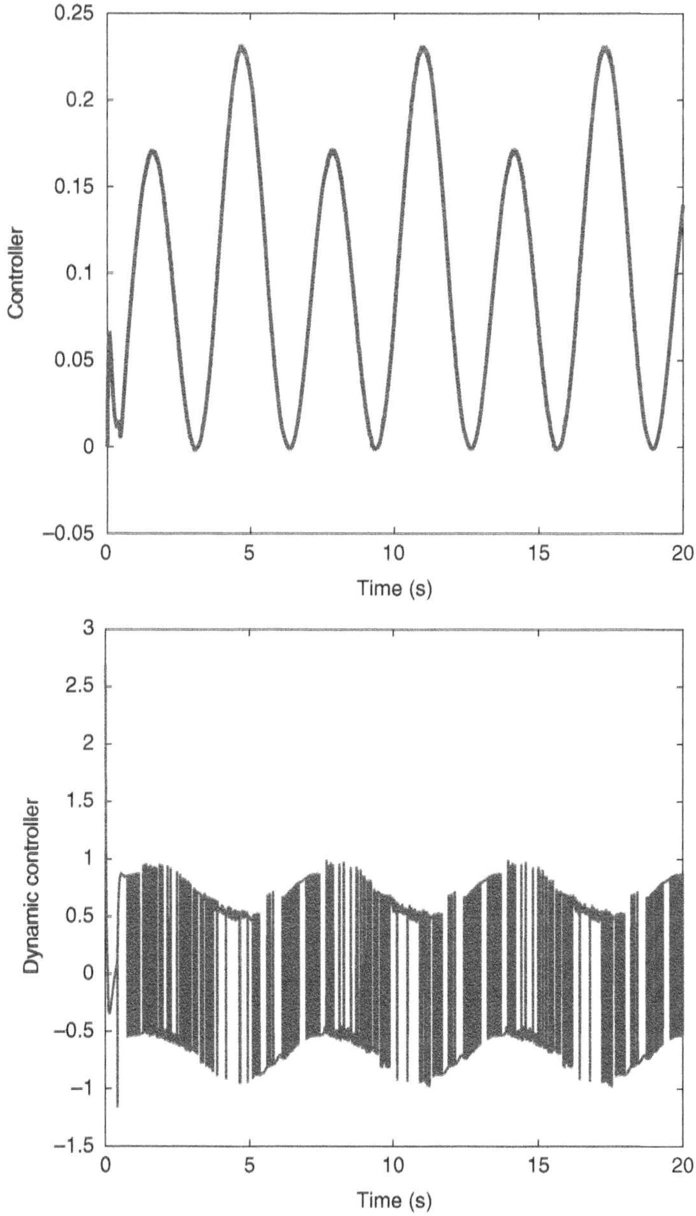

Figure 4.6 Dynamic controller execution in Simulink.

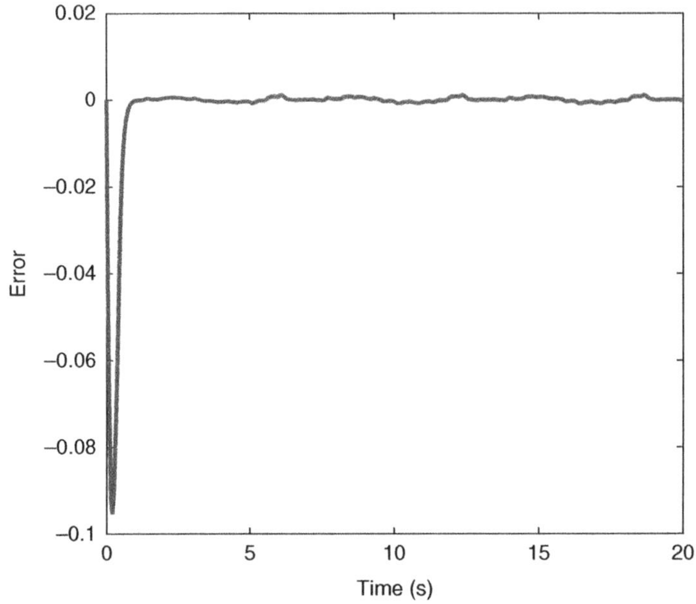

Figure 4.7 Tracking error resulting from program execution in Simulink.

References

1 Koshkouei, A.J., Burnham, K.J., and Zinober, A.S.I. (2005). Dynamic sliding mode control design. *IEE Proceedings: Control Theory and Applications* 152 (4): 392–396.

2 Perruquetti, W. and Barbot, J.-P. (2002). *Sliding Mode Control in Engineering*. CRC Press.

3 Xu, J., Wang, M., and Qiao, L. (2015). Dynamical sliding mode control for the trajectory tracking of underactuated unmanned underwater vehicles. *Ocean Engineering* 105: 54–63.

5

Fuzzy Sliding Mode Controllers

5.1 Introduction

As stated in the previous sections, many methods have been presented to deal with the limitations of sliding mode controllers, such as chattering or unwanted fluctuations, and improving the convergence speed. One of these methods is fuzzy sliding mode controller (FSMC), which has the advantages of both sliding mode and fuzzy control methods. In addition to the robustness improvement, the chattering is significantly reduced using FSMC

5.2 Designing Adaptive Fuzzy Sliding Mode Controller

To design the controller the universal approximation theorem is considered.

Theorem 5.1 Universal Approximation Theorem
Consider U as a close set in R^n. Then for every continuous function $H(x)$ in U and an arbitrary $\varepsilon > 0$ there exists a fuzzy system $M(x)$ such that

$$\sup_{x \in U} |M(x) - H(x)| < \varepsilon \tag{5.1}$$

It means that fuzzy systems including singleton fuzzifiers, center average de-fuzzifier, product inference engine, and Gaussian membership functions are universal approximators, and the proof is proposed in [1] based on Stone–Weierstrass Theorem.

Methods of Developing Sliding Mode Controllers: Design and MATLAB Simulation, First Edition.
Reihaneh Kardehi Moghaddam and Mostafa Rabbani.

Published 2025 by John Wiley & Sons, Inc.

The above theorem shows that a certain class of nonlinear functions can be approximated by the fuzzy system with desired accuracy. In the following, an adaptive fuzzy sliding mode control design algorithm is introduced for a class of non-deterministic nonlinear systems, and then it is implemented in design of control system.

The non-deterministic nonlinearities are approximated by the fuzzy system, with a set of if-then rules whose parameters are adjusted online according to the adaptation rules. This method helps to control the output of the nonlinear system to follow a certain path. The Lyapunov combination method is used to develop an adaptive control algorithm based on the adaptive fuzzy model. The chattering phenomenon is addressed and the robustness is guaranteed. In addition, the stability analysis of the proposed control method is performed.

5.2.1 Designing Sliding Mode Controller Based on Fuzzy Approximator

Consider the following nonlinear system, where f and g are nonlinear functions, $u \in R$ is control input, and $|d(t)| \leq D$ is the disturbance.

$$\ddot{x} = f(x,\dot{x}) + g(x,\dot{x})u + d(t) \tag{5.2}$$

x_d is considered to be the desired path as follows:

$$e = x_d - x \tag{5.3}$$

Considering the tracking error of (5.3), sliding surface is defined as:

$$s = \dot{e} + ce \tag{5.4}$$

considering $c > 0$, $\dot{s}$ is obtained as:

$$\dot{s} = \ddot{e} + c\dot{e} = \ddot{x}_d - \ddot{x} + c\dot{e} = \ddot{x}_d - f - gu - d(t) + c\dot{e} \tag{5.5}$$

If f and g are considered to be known, then controlling law is easily obtained as follows:

$$u = g^{-1}[-f + \ddot{x}_d + c\dot{e} + \eta\,\mathrm{sign}(s)] \tag{5.6}$$

Substituting (5.6) in (5.5) yields:

$$\dot{s} = \ddot{x}_d - f - gu - d(t) + c\dot{e} = -\eta\,\mathrm{sign}(s) - d(t) \tag{5.7}$$

Defining $\eta \geq D$, yields:

$$s\dot{s} = -\eta|s| - s.d(t) \leq 0 \tag{5.8}$$

It is concluded that the closed-loop system is stable according to Lyapunov stability theorem. If the system is not affected by uncertainties, then model (5.2) and control (5.6) will be reliable. Here, it is assumed that the exact value of the function f is unknown, and it is proposed to use fuzzy approximation instead.

5.2.2 Fuzzy Approximation

Based on Theorem 5.1, $f(x)$ can be approximated by $\hat{f}(x)$. To utilize universal approximation theorem, the following steps are considered:

Step 1: For $x_i(i = 1, 2, ..., n)$, fuzzy sets are denoted as $A_i^{l_i}$ $l_i = 1, 2, ..., p_i$

Step 2: $\prod_{i=1}^{n} p_i$ number of fuzzy rules are considered to construct $\hat{f}(x \mid \theta_f)$

where $l_i = 1, 2, ..., p_i$ and $i = 1, 2, ..., n$

$$R^{(j)}: \text{If } x_1 \text{ is } A_1^{l_1} \text{ and ... and } x_n \text{ is } A_l^{l_n} \text{ then } \hat{f} \text{ is } E^{l_1,...,l_n} \tag{5.9}$$

Note 5.1

Since the fuzzy system $\hat{f}(x)$ is not able to approximate $f(x)$ sufficiently, in order to increase the accuracy of $\hat{f}(x)$, some parameters of $\hat{f}(x)$ are considered to be free. They are able to change during online operations, so that the accuracy of the approximation improves over time.

The fuzzy system output is [2]:

$$\hat{f}(x|\theta_f) = \frac{\sum_{l_1=1}^{p_1} \cdots \sum_{l_n=1}^{p_n} \bar{y}_f^{l_1,...,l_n} \left(\prod_{i=1}^{n} \mu_{A_i^{l_i}}(x) \right)}{\sum_{l_1=1}^{p_1} \cdots \sum_{l_n=1}^{p_n} \left(\prod_{i=1}^{n} \mu_{A_i^{l_i}}(x) \right)} \tag{5.10}$$

in which $\mu_{A_i^{l_i}}(x_i)$ is the membership value of xi related to $A_i^{l_i}$

Let $\hat{\theta}f = [\bar{y}_f^{l_1,...,l_n}]$ ($\hat{\boldsymbol{\theta}}_f \in \mathbf{R}^{\prod_{i=1}^{n} p_i}$) be the parameter that must be approximated, and

$$\xi_{l_1,...,l_n}(x) = \frac{\prod_{i=1}^{n} \mu_{A_i^{l_i}}(x)}{\sum_{l_1=1}^{p_1} \cdots \sum_{l_n=1}^{p_n} \left(\prod_{i=1}^{n} \mu_{A_i^{l_i}}(x) \right)} \tag{5.11}$$

Then (5.10) is rewritten as:

$$\hat{f}(x|\theta_f) = \hat{\theta}_f^T \xi(x) \tag{5.12}$$

The fuzzy membership function is selected by expert, and clearly all the states must be measurable and known.

Based on Theorem 5.1, a set of optimal parameters exist that result in minimizing the approximation error.

$$f = \theta_f^{*T}\xi(x) + \varepsilon \tag{5.13}$$

where x is the least approximation error. Thus, the total approximation error is [2]

$$\tilde{f} = f - \hat{f} = \theta_f^{*T}\xi(x) + \varepsilon - \hat{\theta}_f^T\xi(x) \tag{5.14}$$

5.3 Designing an Indirect Adaptive Fuzzy Sliding Mode Controller

Adaptive control is generally divided into two types of indirect and direct adaptive controlling methods. In the indirect comparative method, first the process model is obtained and then a controller based on the model is proposed according to the obtained model. As a result, considering the fact that there is no actual model because it does not actually have the model, it uses the approximation of the model in the controller and adapts it. Indirect adaptive control is based on system knowledge [3].

In the direct adaptive method, a controller is used without considering the process model, and then the controller parameters are adapted according to the efficiency and the created error. Direct adaptive control is basically based on control knowledge.

The optimal parameter is defined as follows [4, 5]:

$$\theta_f^* = \arg\min_{\theta_f \in \Omega_f}\left(\sup_{x \in R^n}\left|\hat{f}(x|\theta_f) - f(x)\right|\right) \tag{5.15}$$

where $\theta_f \epsilon \Omega_f$

f is proposed as:

$$f = \theta_f^{*T}\xi(x) + \varepsilon \tag{5.16}$$

where x is the input signal of the fuzzy system, $\xi(x)$ is the fuzzy vector, and ε is the approximation error of the fuzzy system, and $\varepsilon \le \varepsilon_N$.

Fuzzy system is used to approximate f. The input of the fuzzy system is selected as $x = [e \quad \dot{e}]^T$ and the output of the fuzzy system is obtained from Eq. (5.12).

The control input is written as:

$$u = g^{-1}\left[-\hat{f} + \ddot{x}_d + c\dot{e} + \eta\,\mathrm{sign}(s)\right] \tag{5.17}$$

Substituting (5.17) in (5.7) yields:

$$\dot{s} = -f + \hat{f} - \eta\,\mathrm{sign}(s) - d(t) = -\tilde{f} - d(t) - \eta\,\mathrm{sign}(s) \tag{5.18}$$

where

$$\tilde{f} = f - \hat{f} = \theta_f^{*T}\xi(x) + \varepsilon - \hat{\theta}_f^T\xi(x) = \tilde{\theta}_f^T\xi(x) + \varepsilon \text{ and } \tilde{\theta}_f = \theta_f^* - \hat{\theta}_f$$

Consider the Lyapunov candidate as follows:

$$V = \frac{1}{2}s^2 + \frac{1}{2}\gamma\tilde{\theta}_f^T\theta_f;\ \gamma > 0 \tag{5.19}$$

Taking the derivative of V and considering (5.18) leads to:

$$\begin{aligned}\dot{V} &= s\dot{s} + \gamma\tilde{\theta}_f^T\dot{\tilde{\theta}}_f = s\left(-\tilde{f} - d(t) - \eta\,\mathrm{sign}(s)\right) - \gamma\tilde{\theta}_f^T\dot{\hat{\theta}}_f \\ &= s\left(-\tilde{\theta}_f^T\xi(x) - \varepsilon - d(t) - \eta\,\mathrm{sign}(s)\right) - \gamma\tilde{\theta}_f^T\dot{\hat{\theta}}_f \\ &= -\tilde{\theta}_f^T\left(s\xi(x) + \gamma\dot{\hat{\theta}}_f\right) - s(\varepsilon + d(t) + \eta\,\mathrm{sign}(s))\end{aligned} \tag{5.20}$$

To minimize V, the adaptation law should be chosen such that $\dot{V}$ becomes negative. Accordingly the first term in Eq. (5.20) is considered to be zero, so the adaptation rule is chosen as follows:

$$\dot{\hat{\theta}}_f = -\frac{1}{\gamma}s\xi(x) \tag{5.21}$$

Using (5.21), $\dot{V}$ is rewritten as:

$$\dot{V} = -s(\varepsilon + d(t) + \eta\,\mathrm{sign}(s)) = -s(\varepsilon + d(t)) - \eta|s| \tag{5.22}$$

Suppose $\eta \geq \varepsilon_N + D$, then (5.22) yields $\dot{V} \leq 0$

5.3.1 Fuzzy Adaptive SMC Simulation Results

To simulate the fuzzy adaptive SMC, we consider the inverted pendulum with Eq. (5.23) and f is considered unknown.

$$\begin{aligned}\dot{x}_1 &= x_2 \\ \dot{x}_2 &= f(x) + g(x)u\end{aligned} \tag{5.23}$$

The nonlinear functions $f(x)$ and $g(x)$ are in the form of Eq. (5.24). The control problem is a tracking type. The goal is to follow the desired reference signal by the state variables $[x_1, x_2]$. In this problem, the states of the system are $x = [q, \dot{q}]$.

$$\begin{cases} f(x_1, x_2) = \dfrac{(M+m)g\sin x_1 - mLx_2^2(\sin x_1 \cos x_1)}{\frac{4}{3}(M+m)L - mL\cos^2 x_1} \\ g(x_1, x_2) = \dfrac{\cos x_1}{\frac{4}{3}(M+m)L - mL\cos^2 x_1} \end{cases} \tag{5.24}$$

In this section, we suppose $x_0 = [0.2, 0.1]$ and $x_d = \frac{\pi}{30}\sin(t)$. x_0 is the initial condition and x_d is the input reference signal.

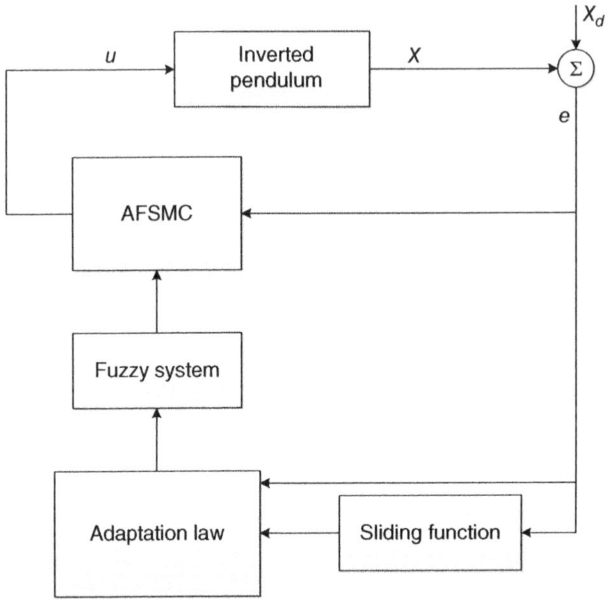

Figure 5.1 Adaptive fuzzy sliding mode control of inverted pendulum.

The parameter c in Eq. (5.4) is 25. Parameters related to inverted pendulum equations are initialized in the program. The controlling structure is defined in Figure 5.1. According to previous section the controller is obtained as follows:

$$\hat{f}_{1\times 1} = \hat{\theta}^T_{f(1\times m)} \xi_{m\times 1} \tag{5.25}$$

$$u = g^{-1}\left[-\hat{f} + \ddot{X}_d + \dot{e} + ce + K\,\mathrm{sign}(s)\right] \tag{5.26}$$

where s is defined for a second-order pendulum system as follows:

$$K\,\mathrm{sign}(s) = K\,\mathrm{sign}(\dot{e} + ce) = \frac{K}{|\dot{e} + ce|}(\dot{e} + ce) = k_d\dot{e} + k_p e \tag{5.27}$$

In the above equation k_p and k_d are considered to be $k_d = \dfrac{K}{|\dot{e} + ce|}$ and $k_p = \dfrac{cK}{|\dot{e} + ce|}$.

Gain k is set to 0.1 in the simulations.

Utilizing controller (5.26) to system (5.23) yields:

$$\ddot{e} + k_d\dot{e} + k_p e + \tilde{f} = 0 \tag{5.28}$$

The simulation of the above system by the Simulink section in MATLAB software is shown in Figure 5.2. S_function blocks are used in the programs of this section.

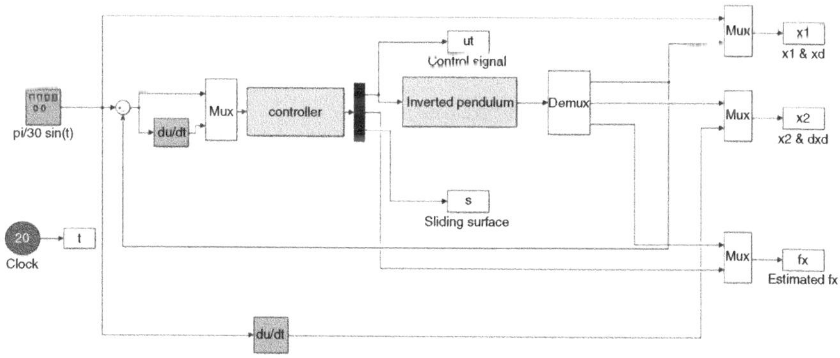

Figure 5.2 Block diagram of the adaptive fuzzy SMC system in Simulink.

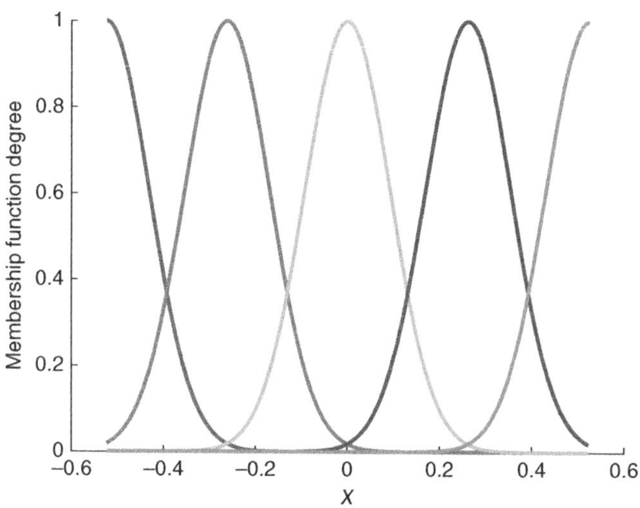

Figure 5.3 The fuzzy membership function.

Fuzzy membership functions are shown in Figure 5.3. The fuzzy program named Fuzzy.m is as follows:

```
clear all;
close all;

L1 = -pi/6;
L2 = pi/6;
L = L2-L1;
T = L*1/1000;
x = L1:T:L2;
```

```
figure(1);

for i = 1:1:5
   gs = -[(x+pi/6-(i-1)*pi/12)/(pi/24)].^2;
    u = exp(gs);
    hold on;
    plot(x,u,'linewidth',3);
end

xlabel('x','FontSize',12,'FontWeight','bold','Color','k');
ylabel('Membership function
degree','FontSize',12,'FontWeight','bold','Color','k');
```

The parameters of the inverted pendulum are given in Table 5.1.

The adaptive parameter in the simulations is $\gamma = 0.05$.

The MATLAB program is defined as follows:

```
function [sys,x0,str,ts] = s_function(t,x,u,flag)
switch flag,
case 0,
    [sys,x0,str,ts] = mdlInitializeSizes;
case 1,
    sys = mdlDerivatives(t,x,u);
case 3,
    sys = mdlOutputs(t,x,u);
case {2, 4, 9 }
    sys = [];
otherwise
    error(['Unhandled flag = ',num2str(flag)]);
end
function [sys,x0,str,ts] = mdlInitializeSizes
sizes = simsizes;
```

Table 5.1 System parameters of inverted pendulum.

Symbol	Parameter	Parameter value
M	Cart mass	1 kg
m	Pendulum mass	0.1 kg
g	Acceleration of gravity	9.8 m/s^2
L	Half-length of the pendulum	0.5 m

```
sizes.NumContStates  = 2;
sizes.NumDiscStates  = 0;
sizes.NumOutputs     = 3;
sizes.NumInputs      = 1;
sizes.DirFeedthrough = 0;
sizes.NumSampleTimes = 0;
sys=simsizes(sizes);
x0 =[0.2 0.1];
str = [];
ts = [];
function sys = mdlDerivatives(t,x,u)
m = 0.1;M = 1;L = 0.5;g = 9.8;
fx = ((M+m)*g*sin(x(1))-m*L*(x(2)^2)*sin(x(1))*cos(x
(1)))/((4/3)*(M+m)*L-m*L*(0.5+0.5*cos(2*x(1))));

gx = cos(x(1))/((4/3)*(M+m)*L-m*L*(0.5+0.5*cos(2*x(1))));

sys(1) = x(2);

sys(2) = fx+gx*u;

function sys = mdlOutputs(t,x,u)
m = 0.1;
M = 1;
L = 0.5;g = 9.8;
fx = ((M+m)*g*sin(x(1))-m*L*(x(2)^2)*sin(x(1))*cos(x
(1)))/((4/3)*(M+m)*L-m*L*(0.5+0.5*cos(2*x(1))));
sys(1) = x(1);
sys(2) = x(2);
sys(3) = fx;
```

The FSMC is designed as follows:

```
function [sys,x0,str,ts] = spacemodel(t,x,u,flag)
switch flag,
case 0,
    [sys,x0,str,ts] = mdlInitializeSizes;
case 1,
    sys = mdlDerivatives(t,x,u);
case 3,
    sys = mdlOutputs(t,x,u);
case {2,4,9}
    sys = [];
```

```
otherwise
    error(['Unhandled flag = ',num2str(flag)]);
end
function [sys,x0,str,ts] = mdlInitializeSizes
sizes = simsizes;
sizes.NumContStates  = 25;
sizes.NumDiscStates  = 0;
sizes.NumOutputs     = 3;
sizes.NumInputs      = 2;
sizes.DirFeedthrough = 1;
sizes.NumSampleTimes = 0;
sys = simsizes(sizes);
x0  = [zeros(25,1)];
str = [];
ts  = [];
function sys = mdlDerivatives(t,x,u)
gama = 0.005;
xd = pi/30*sin(t); dxd = pi/30*cos(t); ddxd = -pi/30*sin(t);

e = u(1); de = u(2);
c = 25;
s = c*e+de;
x1 = e; x2 = de;

for i = 1:1:25
    thtaf(i,1) = x(i);
end

FS1 = 0;
for l1 = 1:1:5
   gs1 = -[(x1+pi/6-(l1-1)*pi/12)/(pi/24)]^2;
    u1(l1) = exp(gs1);
end

for l2 = 1:1:5
   gs2 = -[(x2+pi/6-(l2-1)*pi/12)/(pi/24)]^2;
    u2(l2) = exp(gs2);
end
for l1 = 1:1:5
    for l2 = 1:1:5
        FS2(5*(l1-1)+l2) = u1(l1)*u2(l2);
        FS1 = FS1+u1(l1)*u2(l2);
```

```
    end
end
FS = FS2/FS1;
for i = 1:1:25
    sys(i) = -1/gama*s*FS(i);
end
function sys = mdlOutputs(t,x,u)
xd = pi/30*sin(t); dxd = pi/30*cos(t); ddxd = -pi/30*sin(t);
e = u(1); de = u(2);
c = 25; s = c*e+de;
x1 = e; x2 = de;
for i = 1:1:25
    thtaf(i,1) = x(i);
end
FS1 = 0;
for l1 = 1:1:5
   gs1 = -[(x1+pi/6-(l1-1)*pi/12)/(pi/24)]^2;
    u1(l1) = exp(gs1);
end
for l2 = 1:1:5
   gs2 = -[(x2+pi/6-(l2-1)*pi/12)/(pi/24)]^2;
    u2(l2) = exp(gs2);
end
for l1 = 1:1:5
    for l2 = 1:1:5
        FS2(5*(l1-1)+l2) = u1(l1)*u2(l2);
        FS1 = FS1+u1(l1)*u2(l2);
    end
end

FS = FS2/FS1;
fxp = thtaf'*FS';

m = 0.1;M = 1;L = 0.5;g = 9.8;
gx = cos(x1)/((4/3)*(M+m)*L-m*L*(0.5+0.5*cos(2*x1)));
k = 0.1;
m = k*s/(abs(s));   %k*sign(s)

ut = 1/gx*(-fxp+ddxd+c*de+m);

sys(1) = ut;sys(2) = fxp;sys(3) = s;
```

The results are plotted using the following instructions:

```
close all;

figure(1);
subplot(211);
plot(t,x1(:,1),'b',t,x1(:,2),'r--','linewidth',3);
xlabel('time(s)','FontSize',12,'FontWeight','bold',
'Color','k');
ylabel('X1 state change in pursuit of desired input
value');
legend('Xd','Fuzzy SMC');

subplot(212);
plot(t,x1(:,1)-x1(:,2),'linewidth',3);
xlabel('time(s)','FontSize',12,'FontWeight','bold',
'Color','k');
ylabel('error','FontSize',12,'FontWeight','bold',
'Color','k');
legend('error signal');

figure(2)
subplot(211);
plot(t,x2(:,1),'b',t,x2(:,2),'r--','linewidth',3);
xlabel('time(s)','FontSize',12,'FontWeight','bold',
'Color','k');
ylabel('X2 state change in pursuit of desired input value');
legend('dXd','Fuzzy SMC');

subplot(212);
plot(t,x2(:,1)-x2(:,2),'linewidth',3);
xlabel('time(s)','FontSize',12,'FontWeight','bold',
'Color','k');
ylabel('error','FontSize',12,'FontWeight','bold',
'Color','k');
legend('error signal');

figure(3);
plot(t,fx(:,1),'b',t,fx(:,2),'r--','linewidth',3);
xlabel('time(s)','FontSize',12,'FontWeight','bold',
'Color','k');
```

```
ylabel('fx and estimated fx','FontSize',12,'FontWeight',
'bold','Color','k');
legend('fx','estimated fx');

figure(4);
plot(t,ut,'b','linewidth',3);
xlabel('time(s)','FontSize',12,'FontWeight','bold',
'Color','k');
ylabel('control signal','FontSize',12,'FontWeight',
'bold','Color','k');

figure(5);
plot(t,s(:,1),'k','linewidth',3);
xlabel('time(s)','FontSize',12,'FontWeight','bold',
'Color','k');
ylabel('sliding surface','FontSize',12,'FontWeight',
'bold','Color','k');
```

It is considered that the inputs of the fuzzy system are the error and its derivative. Figures 5.4 and 5.5 indicate that after six seconds, the error becomes zero and the state variables follow the desired input signal well.

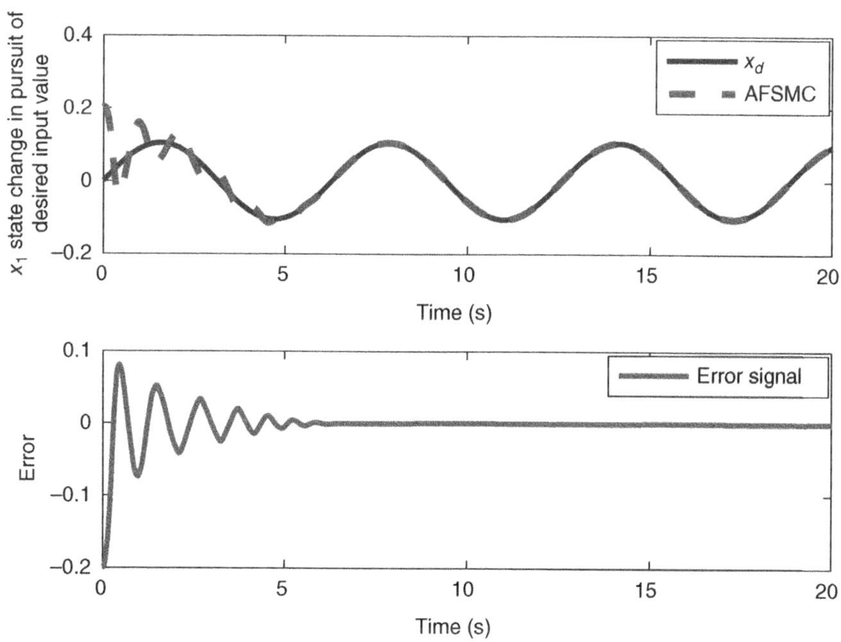

Figure 5.4 Tracking error of x_1 using fuzzy adaptive SMC (FASMC).

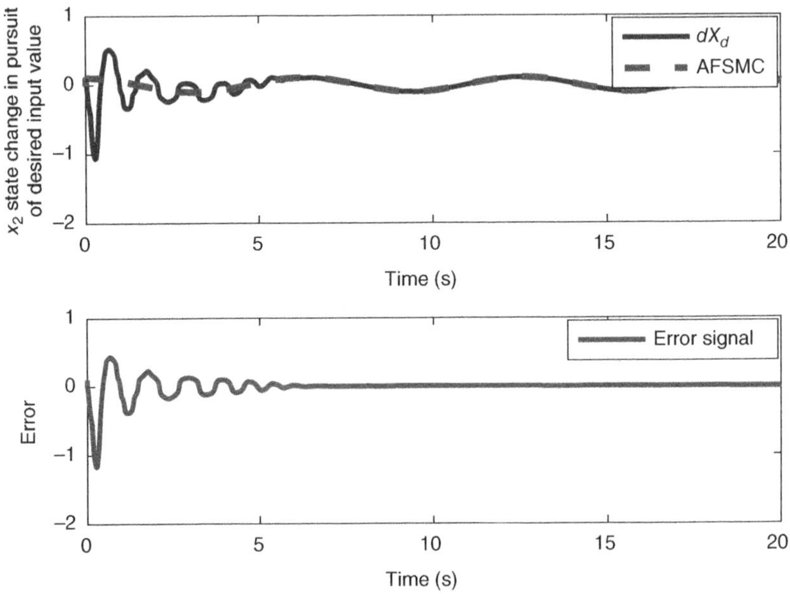

Figure 5.5 Tracking error of x_2 using fuzzy adaptive SMC (FASMC).

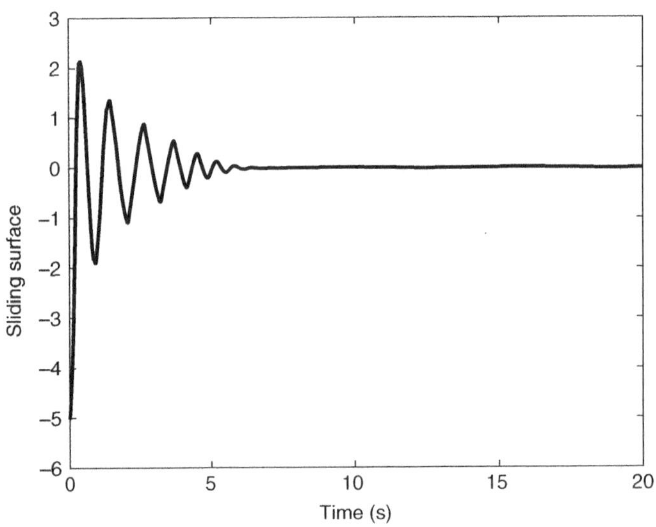

Figure 5.6 Sliding surface in AFSMC.

Exactly after six seconds, the state variables are absorbed in the sliding surface (Figure 5.6); also the chattering is removed from the control signal (Figure 5.7) and the $f(x)$ function is estimated well (Figure 5.8).

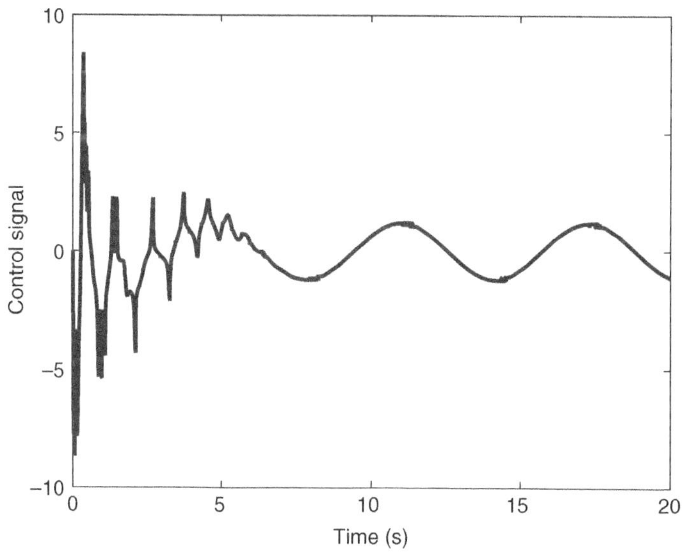

Figure 5.7 Control signal in AFSMC.

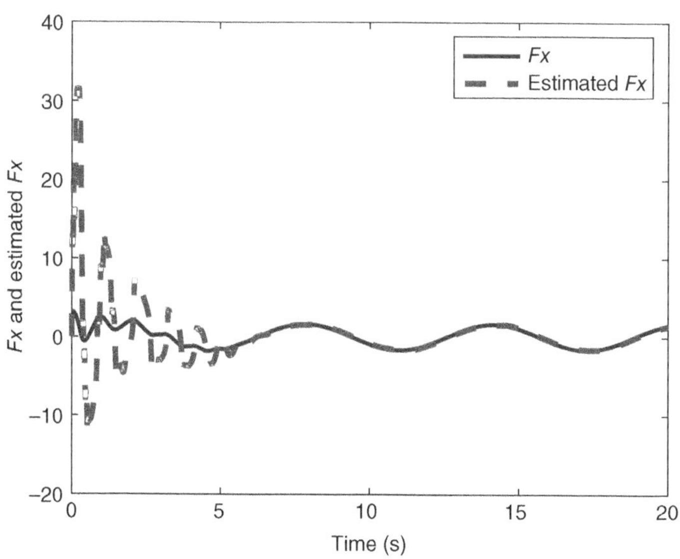

Figure 5.8 Diagram of how to estimate the unknown $f(x)$ function.

5.4 Designing Sliding Mode Controller Based on Fuzzy Equivalent Control

As introduced in the previous chapters, the control law consists of an equivalent control u_{eq} section and a switch control u_{sw} section. In fact, the system-state variables reach the sliding surface with the help of the switch control law and remain on the sliding level with the help of the equivalent control law. With the help of fuzzy rules, the fuzzy system is created based on equivalent and switch control rules [4]. In this section, the switching gain of the control law can be selected with the help of fuzzy rules and fuzzy system. The final control law is introduced in this section as follows:

$$u = u_{eq} - \mu\, u_{sw} \tag{5.29}$$

When $\mu = 1$, it is actually classical sliding mode control. For $\mu \neq 1$, the chattering phenomenon can be reduced by selecting the type of membership function μ in a fuzzy way. In the sliding mode control method based on fuzzy equivalent control, the variable μ is changed according to sliding surface $s(t)$ as follows:

$$\text{If } s(t) \text{ is Negative then } \mu \text{ is Positive} \tag{5.30}$$

$$\text{If } s(t) \text{ is Zero then } \mu \text{ is Zero} \tag{5.31}$$

$$\text{If } s(t) \text{ is Positive then } \mu \text{ is Positive} \tag{5.32}$$

Fuzzy laws (5.30) and (5.32) show that the fuzzy controller is actually the control law (5.29) when the switch function is not zero. Also, fuzzy law (5.31) shows that the fuzzy controller is actually the equivalent controller u_{eq} when the gain of the switch function is zero.

5.4.1 Fuzzy Equivalent SMC Simulation Results for Inverted Pendulum

In this section, the fuzzy controller is simulated on chaotic Duffing and inverted pendulum systems. In the first example, the inverted pendulum equations are as described in (5.23).

Initial states are $x_0 = [0.05, 0.1]$ and $x_d = \sin(t)$. x_0 is the initial condition and x_d is the input reference signal. The parameter c in Eq. (5.4) is chosen equal to 25. Parameters related to inverted pendulum equations are initialized in the program.

The simulation of the above system is shown in Figure 5.9. S_function blocks are used in the programs of this section.

The related results are displayed as (5.13) and (5.14) where the input and out-put membership functions are defined in Figures 5.10 and 5.11. The controller is programmed in the following file:

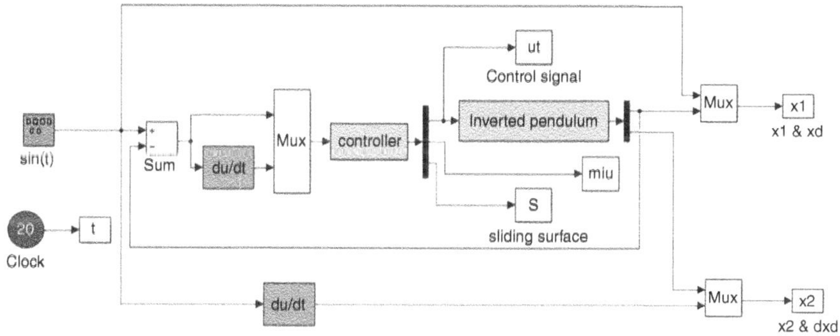

Figure 5.9 Block diagram of the fuzzy equivalent SMC in Simulink.

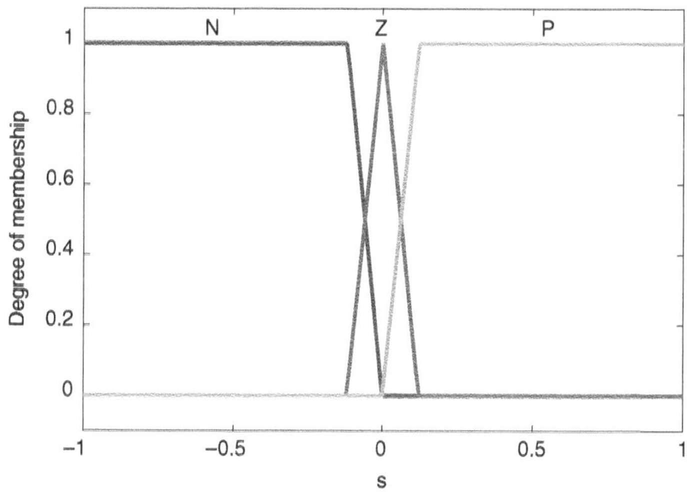

Figure 5.10 The membership function of fuzzy input s.

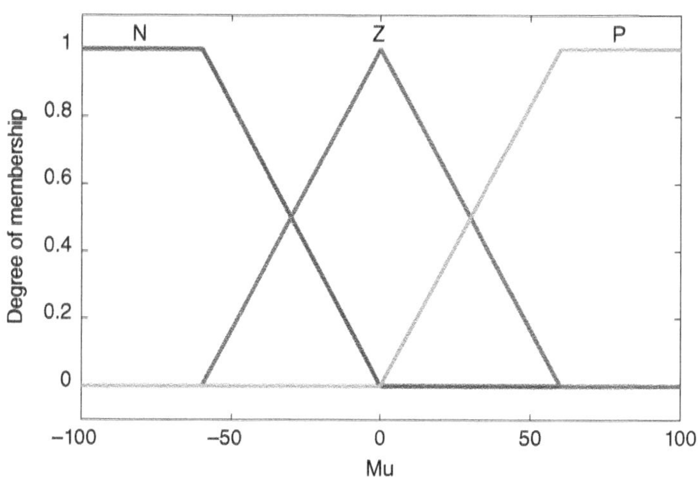

Figure 5.11 The membership function of fuzzy output μ.

```
close all
clear all;

%% input

a = newfis('fuzz_smc');
a = addvar(a,'input','s',1/25*[-25,25]);
a = addmf(a,'input',1,'N','trapmf',1/25*[-25,-25,-3,0]);
a = addmf(a,'input',1,'Z','trimf',1/25*[-3,0,3]);
a = addmf(a,'input',1,'P','trapmf',1/25*[0,3,25,25]);

%% output

a = addvar(a,'output','Mu',20*[-5,5]);
a = addmf(a,'output',1,'N','trapmf',20*[-5,-5,-3,0]);
a = addmf(a,'output',1,'Z','trimf',20*[-3,0,3]);
a = addmf(a,'output',1,'P','trapmf',20*[0,3,5,5]);

%% Create the rule list

rulelist = [1 3 1 1;
            2 2 1 1;
            3 3 1 1];

a = addrule(a,rulelist);
showrule(a)

a1 = setfis(a,'DefuzzMethod','centroid');   %Defuzzy

writefis(a1,'fsmc');        %Save fuzzy system as "fsmc.fis"
a2 = readfis('fsmc');
ruleview(a2);

%% plot
figure(1);
plotmf(a,'input',1);
figure(2);
plotmf(a,'output',1);
```

The display of fuzzy rules is as in Figure 5.12.

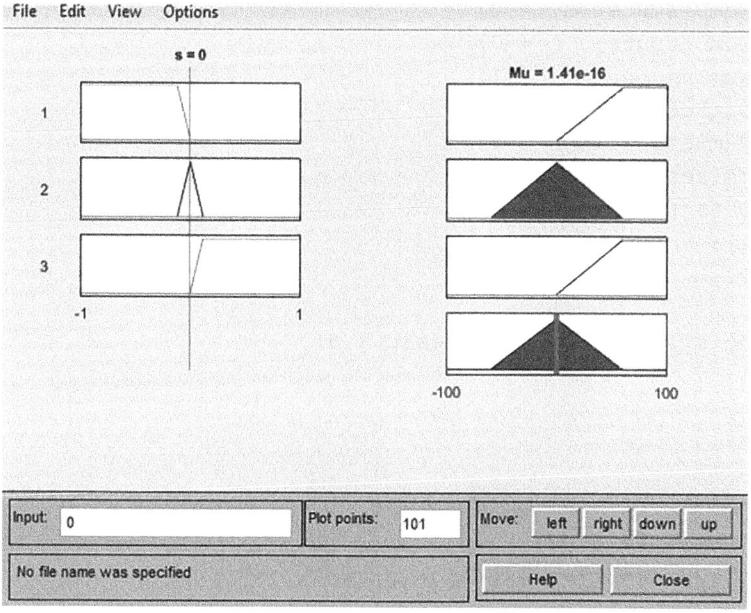

Figure 5.12 Fuzzy rules.

The inverted pendulum dynamics are defined as follows:

```
function [sys,x0,str,ts] = s_function(t,x,u,flag)
switch flag,
case 0,
    [sys,x0,str,ts] = mdlInitializeSizes;
case 1,
    sys = mdlDerivatives(t,x,u);
case 3,
    sys = mdlOutputs(t,x,u);
case {2, 4, 9 }
    sys = [];
otherwise
    error(['Unhandled flag = ',num2str(flag)]);
end

function [sys,x0,str,ts] = mdlInitializeSizes
sizes = simsizes;
sizes.NumContStates  = 2;
```

```
sizes.NumDiscStates   = 0;
sizes.NumOutputs      = 2;
sizes.NumInputs       = 1;
sizes.DirFeedthrough = 0;
sizes.NumSampleTimes = 0;
sys=simsizes(sizes);
x0 = [0.05 0.1];
str = [];
ts = [];

function sys = mdlDerivatives(t,x,u)
m = 0.1;
M = 1;
L = 0.5;
g = 9.8;

fx = ((M+m)*g*sin(x(1))-m*L*(x(2)^2)*sin(x(1))*cos(x
(1)))/((4/3)*(M+m)*L-m*L*(0.5+0.5*cos(2*x(1))));

gx = cos(x(1))/((4/3)*(M+m)*L-m*L*(0.5+0.5*cos(2*x(1))));

sys(1) = x(2);
sys(2) = fx+gx*u;
function sys = mdlOutputs(t,x,u)
sys(1) = x(1);
sys(2) = x(2);
```

The controller is described as below:

```
function [sys,x0,str,ts] = s_function(t,x,u,flag)
switch flag,
case 0,
    [sys,x0,str,ts] = mdlInitializeSizes;
case 3,
    sys = mdlOutputs(t,x,u);
case {2, 4, 9 }
    sys = [];
otherwise
    error(['Unhandled flag = ',num2str(flag)]);
end
function [sys,x0,str,ts] = mdlInitializeSizes
sizes = simsizes;
```

```
sizes.NumContStates  = 0;
sizes.NumDiscStates  = 0;
sizes.NumOutputs     = 3;
sizes.NumInputs      = 2;
sizes.DirFeedthrough = 1;
sizes.NumSampleTimes = 0;
sys=simsizes(sizes);
x0 = [];
str = [];
ts = [];
function sys = mdlOutputs(t,x,u)
persistent a2
if t = = 0
     a2 = readfis('fsmc.fis');
end
xd = sin(t);dxd = cos(t);ddxd = -sin(t);
e = u(1);
de = u(2);
c = 25;
s = c*e+de;
x1 = xd-e;
x2 = dxd-de;
m = 0.1;M = 1;L = 0.5;g = 9.8;
fx = ((M+m)*g*sin(x1)-m*L*(x2^2)*sin(x1)*cos(x1))/((4/3)*
(M+m)*L-m*L*(0.5+0.5*cos(2*x1)));
gx = cos(x1)/((4/3)*(M+m)*L-m*L*(0.5+0.5*cos(2*x1)));

ueq = 1/gx*(c*de+ddxd-fx);
us = 1/gx*sign(s);

w = 2;
if w==1       % Using conventional equivalent sliding mode
control
   Mu = 1.0;
elseif w==2
   Mu = evalfis([s],a2);  % Using fuzzy equivalent sliding
mode control
end
ut = ueq+Mu*us;
sys(1) = ut;
sys(2) = Mu;
sys(3) = s;
```

And finally the results are plotted as follows:

```
close all;
figure(1);

subplot(211);
plot(t,x1(:,1),'b',t,x1(:,2),'r--','linewidth',3);
xlabel('time(s)');ylabel('X1 state change in pursuit of
desired input value');
legend('Xd','Fuzzy SMC');

subplot(212);
plot(t,x1(:,1)-x1(:,2),'linewidth',3);
xlabel('time(s)');ylabel('error');
legend('error signal');

figure(2);
plot(t,ut,'b--','linewidth',3);
xlabel('time(s)');ylabel('control signal');

figure(3);
subplot(211);
plot(t,x2(:,1),'b',t,x2(:,2),'r--','linewidth',3);
xlabel('time(s)');ylabel('X2 state change in pursuit of
desired input value');
legend('dXd','Fuzzy SMC');
subplot(212);
plot(t,x2(:,1)-x2(:,2),'linewidth',3);
xlabel('time(s)');ylabel('error');
legend('error signal');

figure(4);
plot(t,s(:,1),'k--','linewidth',3);
xlabel('time(s)');ylabel('sliding surface');

figure(5);
plot(t,miu(:,1),'c','linewidth',3);
xlabel('time(s)');ylabel('Membership function change');
```

At first, the parameter $\mu = 1$ and the controller is of the classic sliding mode type and the results are as follows.

As can be seen in Figures 5.13 and 5.14, the state variables x_1 and x_2 follow the input reference signal at a good speed and in the very early moments of the simulation run. Also, the error signal suffers from some fluctuations.

In Figure 5.15, the sliding surface converges to zero, and its fluctuation is negligible. In Figure 5.16, the control signal has high fluctuations and chattering is appeared.

For $\mu \neq 1$ the results of reference signal tracking are Figures 5.17 and 5.18 for x_1 and x_2, respectively.

As can be seen in Figures 5.17 and 5.18, the state variables x_1 and x_2 follow the input reference signal at a good speed and in the very early moments of the simulation run. The error is zeroed at the right time.

In the fuzzy sliding method (Figure 5.19), the state variables quickly reach the sliding surface and remain on the surface. The performance is due to the implementation of fuzzy rules on the gain of the switch control function. As can be seen in Figure 5.20, the control signal has oscillation, but the chattering phenomenon has been effectively eliminated.

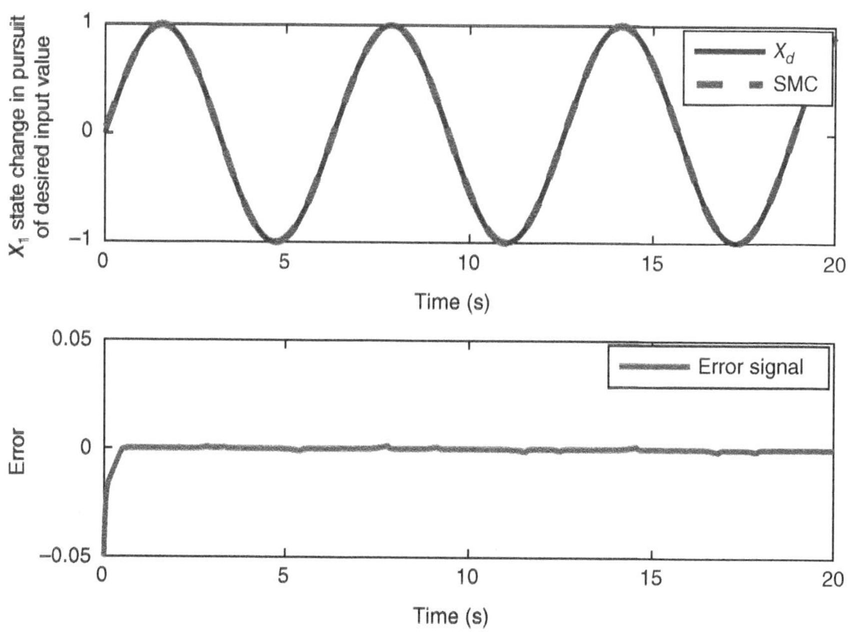

Figure 5.13 Tracking error of x_1 using fuzzy equivalent SMC (FESMC).

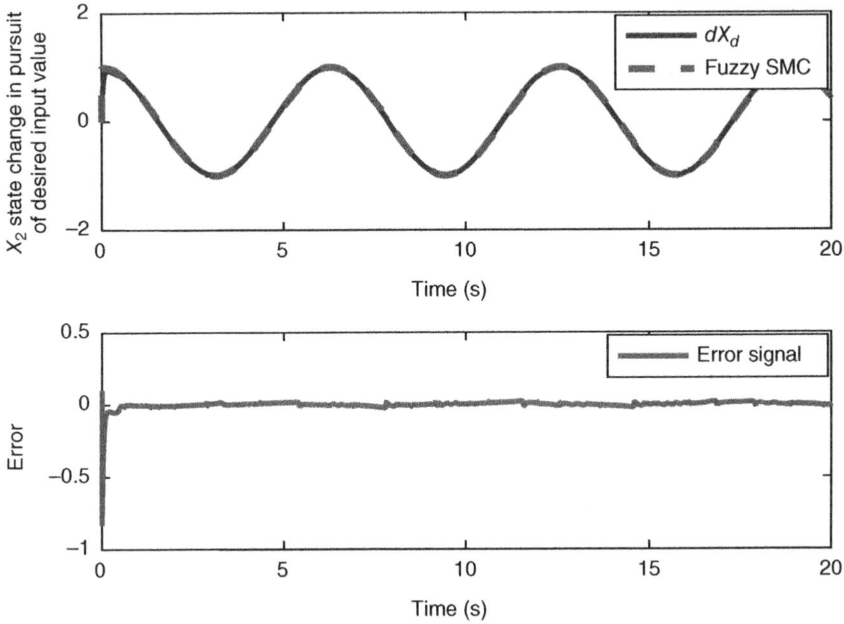

Figure 5.14 Tracking error of x_2 using fuzzy equivalent SMC (FESMC).

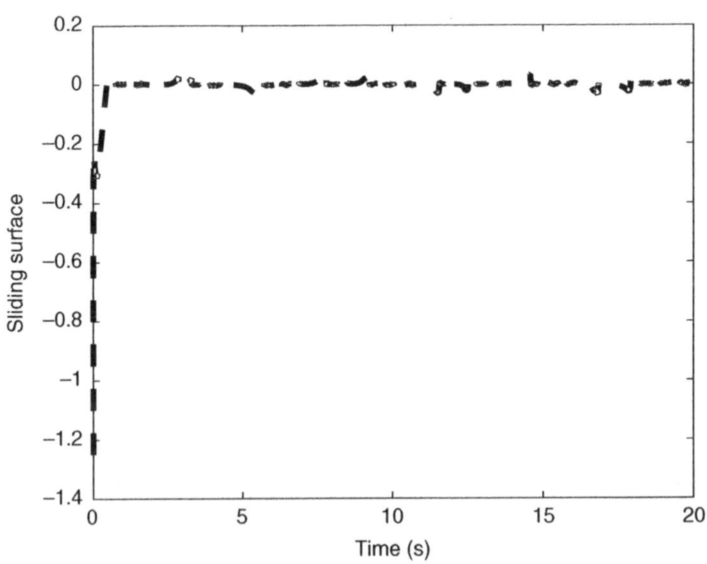

Figure 5.15 Sliding surface in FESMC.

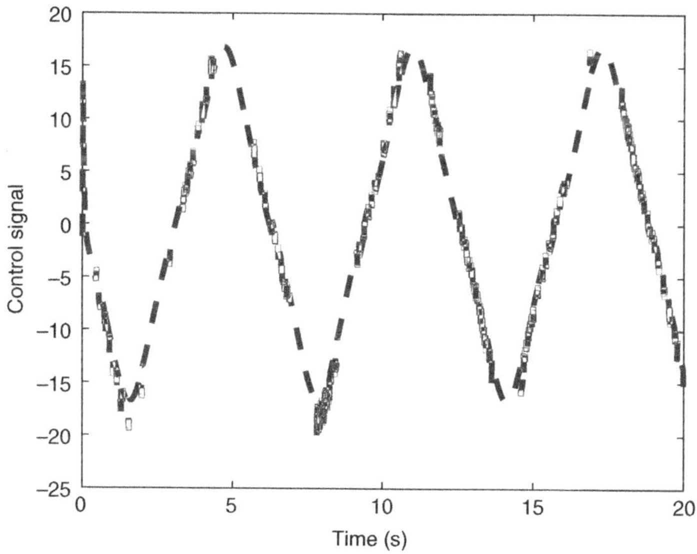

Figure 5.16 Control signal in FESMC.

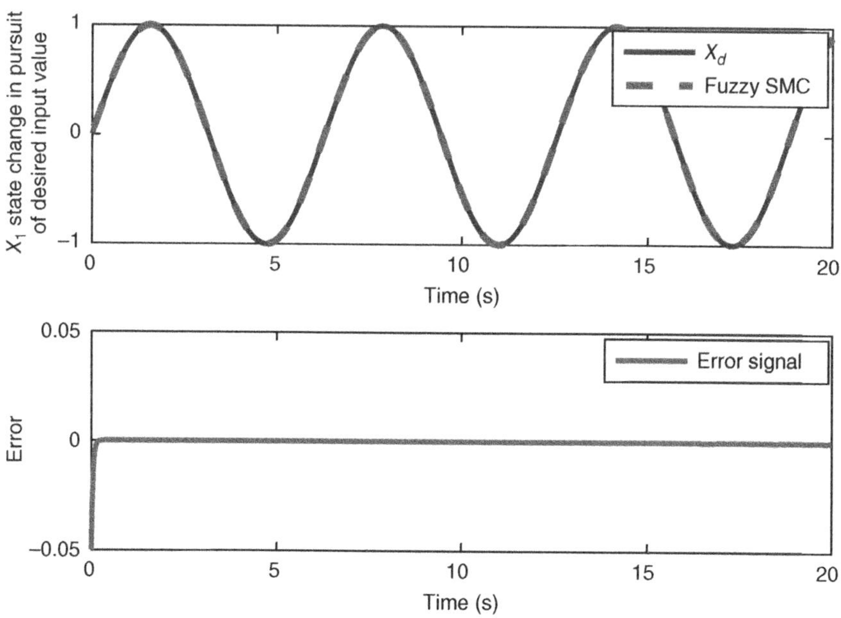

Figure 5.17 x_1 tracking considering $\mu \neq 1$.

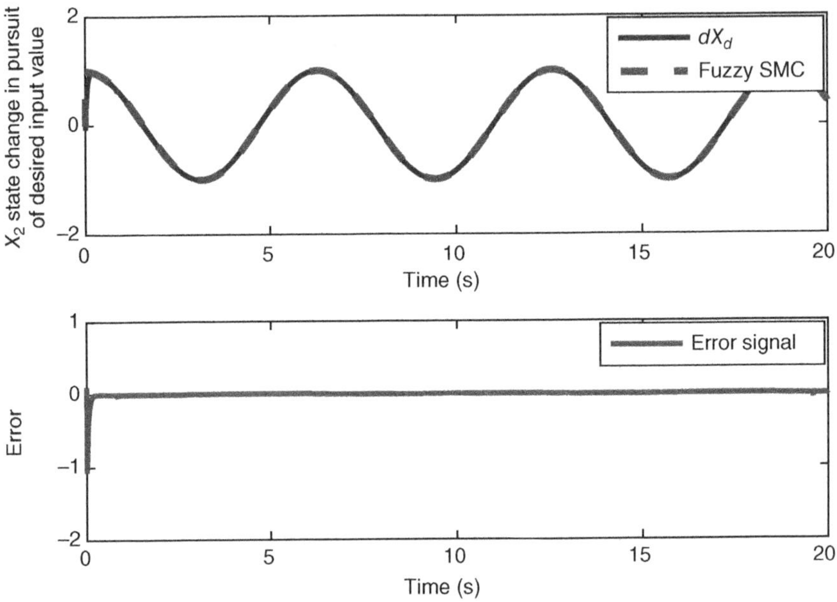

Figure 5.18 x_2 tracking considering $\mu \neq 1$.

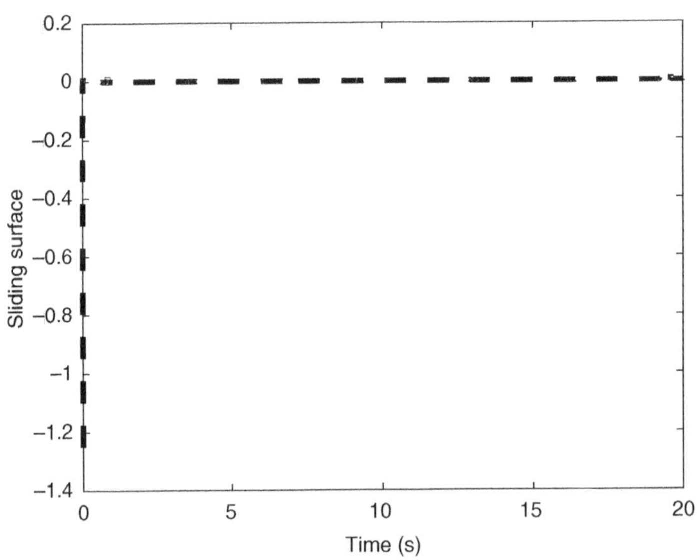

Figure 5.19 Sliding surface in fuzzy SMC.

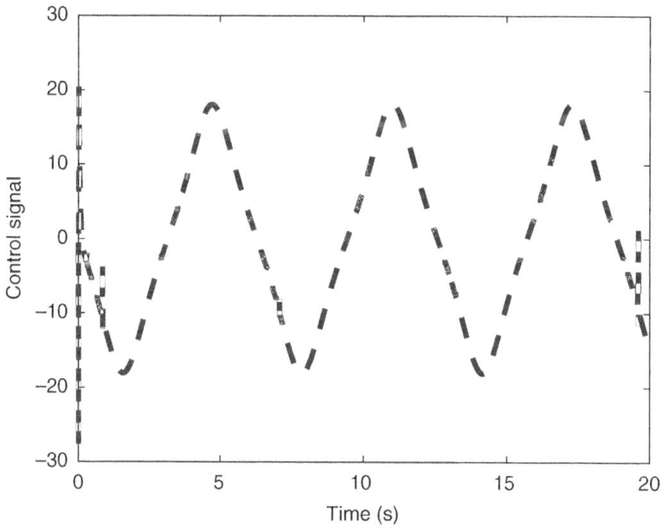

Figure 5.20 Control signal in fuzzy SMC.

5.4.2 Fuzzy Equivalent SMC Simulation Results for Duffing System

Duffing's chaotic system equations are fully introduced in Chapter 8. Consider a second-order chaotic system defined by Duffing equation:

$$\begin{aligned} \dot{x}_1 &= x_2 \\ \dot{x}_2 &= -p_1 x_2 - p_2 x_1 - p_3 x_1^3 + p_4 \cos(\omega t) + u \end{aligned} \tag{5.33}$$

The parameter values of the chaotic Duffing system are as follows:

$$p_1 = 0.4, p_2 = -1.1, p_3 = 1, p_4 = 2.1, \omega = 1.8$$

We suppose $x_0 = [0.05, 0.1]$ and $x_d = \sin(t)$. x_0 is the initial condition and x_d is the input reference signal. The related block diagram is shown in Figure 5.21.

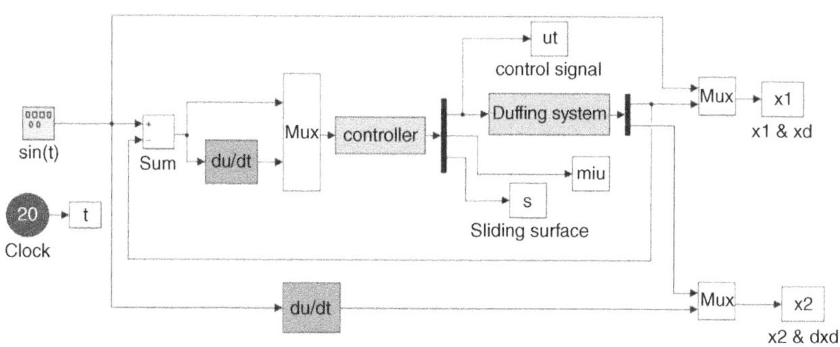

Figure 5.21 Block diagram of the system in Simulink.

To simulate the results the following process is done:

```
function [sys,x0,str,ts] = s_function(t,x,u,flag)
switch flag,
case 0,
    [sys,x0,str,ts] = mdlInitializeSizes;
case 1,
    sys = mdlDerivatives(t,x,u);
case 3,
    sys = mdlOutputs(t,x,u);
case {2, 4, 9 }
    sys = [];
otherwise
    error(['Unhandled flag = ',num2str(flag)]);
end
function [sys,x0,str,ts]=mdlInitializeSizes
sizes = simsizes;
sizes.NumContStates  = 2;
sizes.NumDiscStates  = 0;
sizes.NumOutputs     = 2;
sizes.NumInputs      = 1;
sizes.DirFeedthrough = 0;
sizes.NumSampleTimes = 0;
sys = simsizes(sizes);
x0 = [0.05 0.1];
str = [];
ts = [];
function sys = mdlDerivatives(t,x,u)

global p1 p2 p3 p4 w

p1 = 0.4;
p2 = -1.1;
p3 = 1;
p4 = 2.1;
w = 1.8;

sys(1) = x(2);
fx = -p1*x(2)-p2*x(1)-p3*x(1)^3+p4*cos(w*t);
gx = 1;
sys(2) = fx+gx*u;
```

```
function sys = mdlOutputs(t,x,u)
sys(1) = x(1);
sys(2) = x(2);
```

The FSMC is as follows:

```
function [sys,x0,str,ts] = s_function(t,x,u,flag)
switch flag,
case 0,
    [sys,x0,str,ts] = mdlInitializeSizes;
case 3,
    sys = mdlOutputs(t,x,u);
case {2, 4, 9 }
    sys = [];
otherwise
    error(['Unhandled flag = ',num2str(flag)]);
end
function [sys,x0,str,ts] = mdlInitializeSizes
sizes = simsizes;
sizes.NumContStates  = 0;
sizes.NumDiscStates  = 0;
sizes.NumOutputs     = 3;
sizes.NumInputs      = 2;
sizes.DirFeedthrough = 1;
sizes.NumSampleTimes = 0;
sys=simsizes(sizes);
x0 = []; str = []; ts = [];
function sys = mdlOutputs(t,x,u)
persistent a2
if t==0
    a2 = readfis('fsmc.fis');

end
xd = sin(t); dxd = cos(t); ddxd = -sin(t);
e = u(1); de = u(2);
c = 25;
s = c*e+de;
x1 = xd-e;
x2 = dxd-de;
p1 = 0.4; p2 = -1.1; p3 = 1; p4 = 2.1; w = 1.8;
sys(1) = x2;
fx = -p1*x2-p2*x1-p3*x1^3+p4*cos(w*t);
gx = 1;
ueq = 1/gx*(c*de+ddxd-fx);
```

```
us = 1/gx*sign(s);
w = 2;
if w==1        % Using conventional equivalent sliding mode
control
   Mu = 1.0;
elseif w==2
  Mu = evalfis([s],a2); %Using fuzzy equivalent sliding
mode control
end

ut = ueq+Mu*us;
sys(1) = ut;
sys(2) = Mu;
sys(3) = s;
```

At first, it is considered that $\mu = 1$, so that the controller converts to a classic sliding mode type, and the results are as in Figures 5.22–5.25.

The chattering phenomenon is clearly observed in the control signal; consequently, the FSMC is applied to improve the results, and hence $\mu \neq 1$ is considered; the results are indicated in Figures 5.26–5.29.

The chattering phenomenon was clearly reduced in the fuzzy sliding mode method.

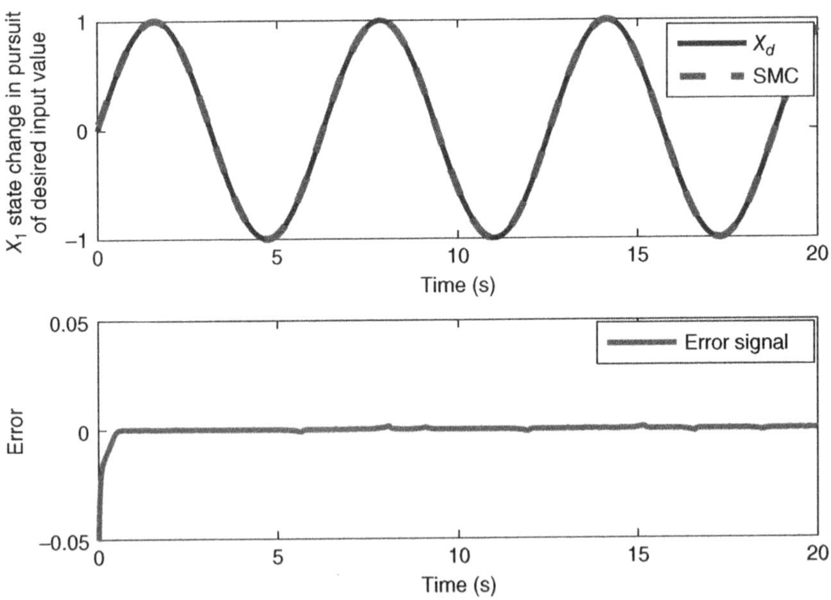

Figure 5.22 x_1 Tracking results for Duffing system using SMC.

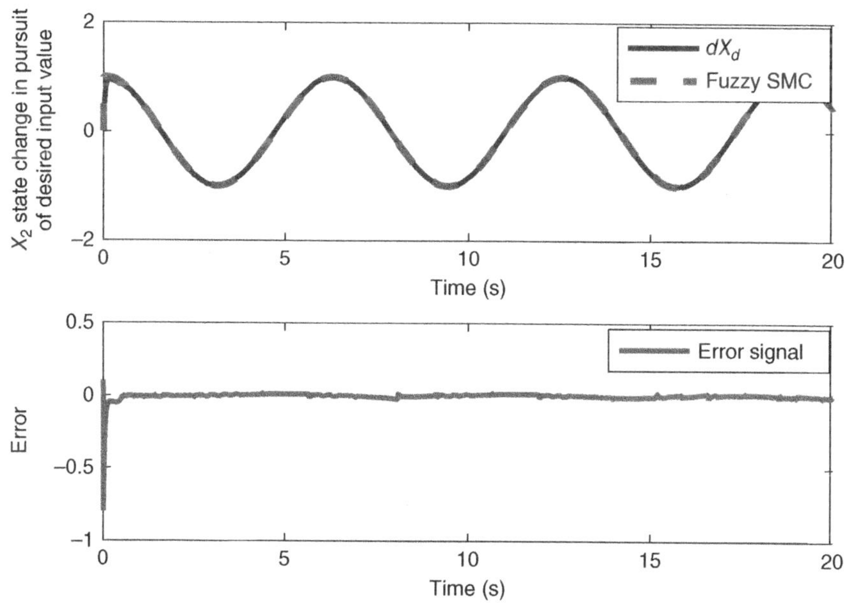

Figure 5.23 x_2 Tracking results for Duffing system using SMC.

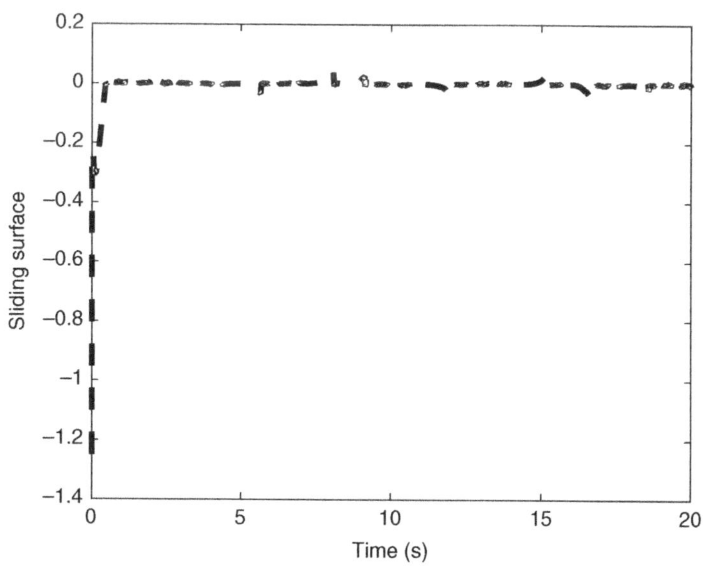

Figure 5.24 Sliding surface in SMC considering μ = 1.

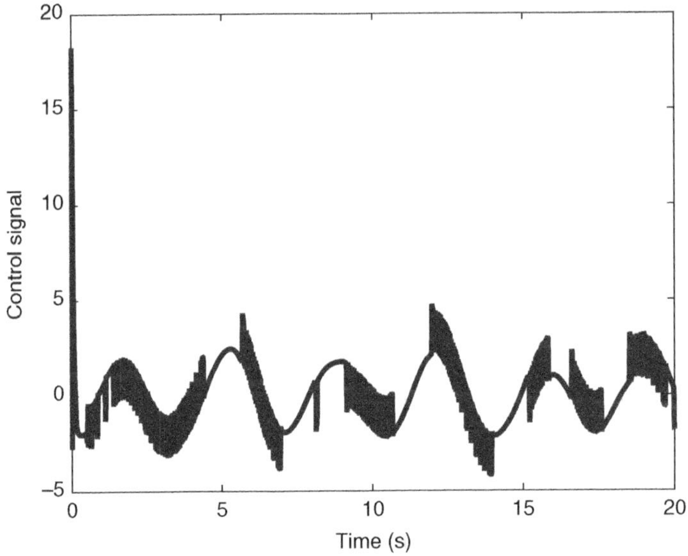

Figure 5.25 Control signal in SMC considering $\mu = 1$.

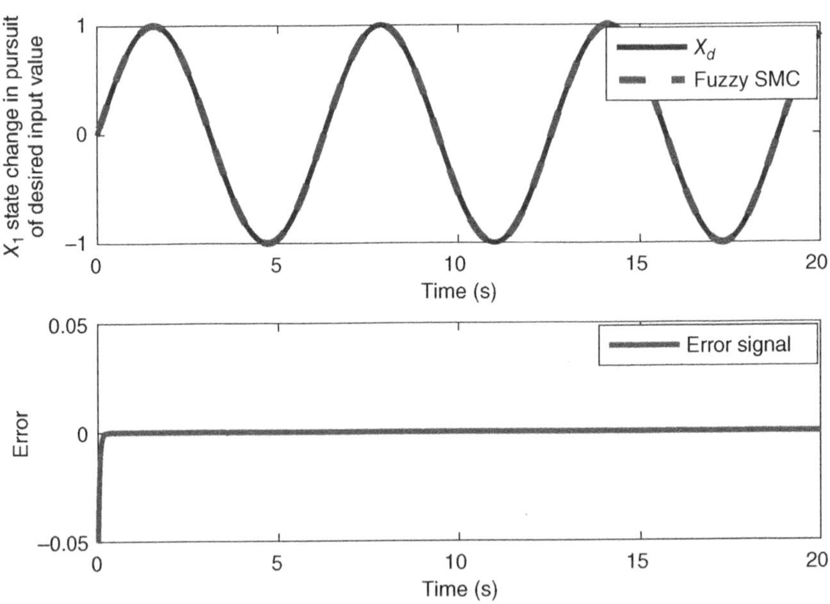

Figure 5.26 x_1 Tracking results for Duffing system using fuzzy SMC.

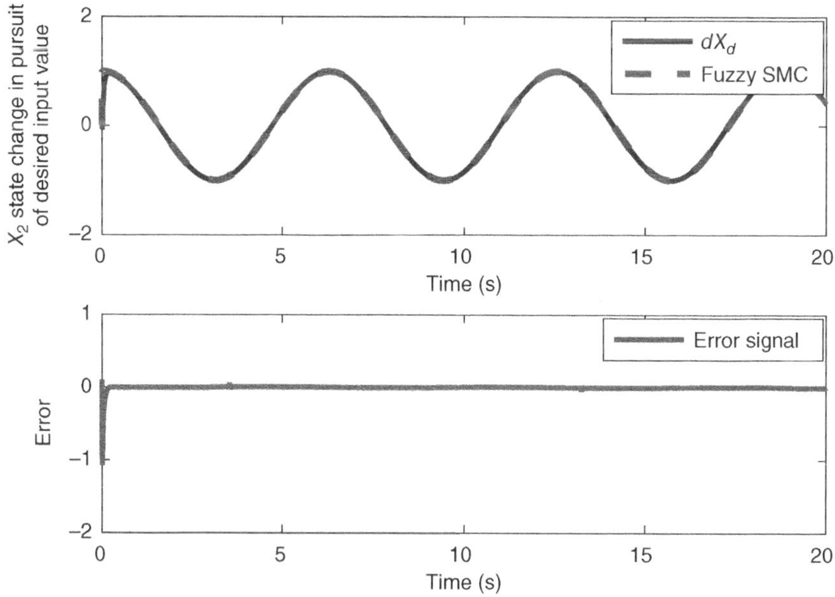

Figure 5.27 x_2 Tracking results for Duffing system using fuzzy SMC.

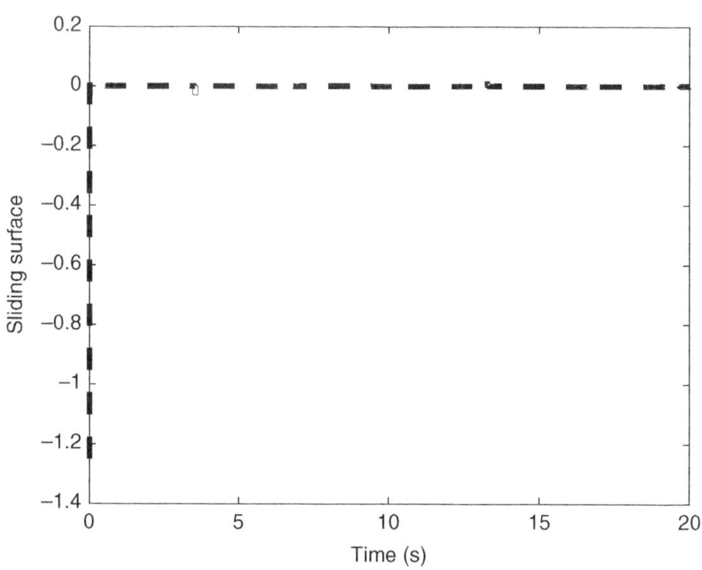

Figure 5.28 Sliding surface of fuzzy SMC for Duffing system.

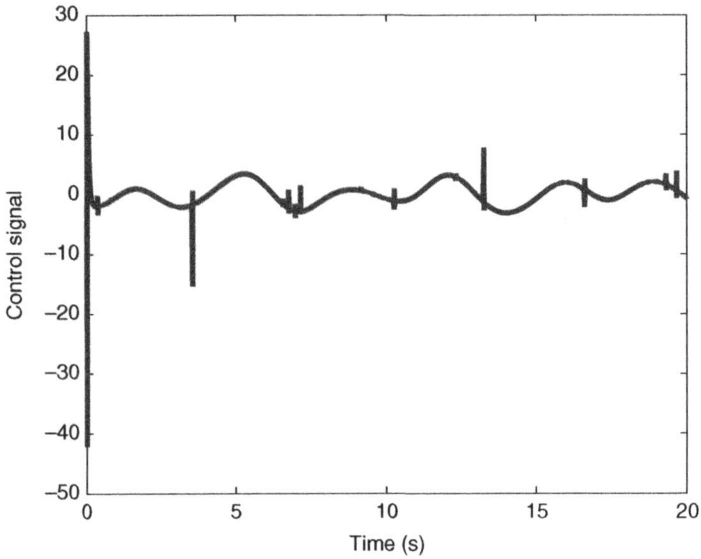

Figure 5.29 Control signal of fuzzy SMC for Duffing system.

References

1 Wang, L.X. (1997). *A Course in Fuzzy System and Control*, 124–149. Prentice Hall.

2 Razzaghian, A. and Kardehi Moghaddam, R. (2016). Adaptive fuzzy sliding mode control for a model-scaled unmanned helicopter. *Journal of Fuzzy Set-Valued Analysis* 3: 286–302. https://doi.org/10.5899/2016/jfsva-00356.

3 Mei, P., Karimi, H.R., Yang, S. et al. (2022). An adaptive fuzzy sliding-mode control for regenerative braking system of electric vehicles. *International Journal of Adaptive Control and Signal Processing* 36, 2: 391–410.

4 Liu, J. and Wang, X. (2012). *Advanced Sliding Mode Control for Mechanical Systems*, 251–260. Springer.

5 Razzaghian, A. and Moghaddam, R.K. (2015). Fuzzy sliding mode control of 5 DOF upper-limb exoskeleton robot. *2015 International Congress on Technology, Communication and Knowledge.*

6

Adaptive Sliding Mode Control

6.1 Introduction

The main limitations of utilizing sliding mode control (SMC) are two interconnected phenomena: chattering and high amplitude of the control action [1]. The larger the magnitude of the discrete controller, the larger chattering amplitude happens.

The mentioned limitations can be handled simultaneously if the magnitude is reduced to a minimal admissible level defined by the conditions for sliding mode to exist. The adaptation methodology for obtaining the minimum possible value of control is proposed as σ-Adaptation, which provides sufficient adjustment for the value of a discontinuous control within the reaching phase [2], and dynamic adaptation [3], which is an adaptation within a sliding mode. In this chapter we focus on the second methodology, and in the sequel design and implementation of an adaptive SMC, an inverted pendulum is proposed in MATLAB.

6.2 Adaptive Sliding Mode Control Design

Consider the following second-order nonlinear system:

$$x^{(n)} = f(x,t) + g(x,t)u + d(t) \tag{6.1}$$

where $u \in R^m$ is the control input and $f \in R^n$, $g \in R^{n \times m}$ are bounded nonlinear vector function of the state vectors. For canonical structures the above equation is rewritten as follows:

$$\begin{cases} \dot{x}_1 = x_2 \\ \quad \dots \\ \dot{x}_n = f(x) + g(x)u \end{cases} \tag{6.2}$$

Methods of Developing Sliding Mode Controllers: Design and MATLAB Simulation, First Edition.
Reihaneh Kardehi Moghaddam and Mostafa Rabbani.

Published 2025 by John Wiley & Sons, Inc.

According to [2] the adaptive sliding mode control for the system above with uncertain parameters is defined as follows.

Theorem 6.1
Consider the nonlinear uncertain system (6.1) with the sliding surface s which is previously defined in (2.2). If the controlling law is considered as (6.3) and (6.4), then there exists a finite time t_f so that a sliding mode is established for all $t \geq t_f$ such that $s = 0$ for all $t \geq t_f$.

$$u = -K\,\mathrm{sign}(s) \tag{6.3}$$

$$K = \begin{cases} \text{the solution of } \dot{K} = K_1|s|; K_1 > 0 & \text{if } s \neq 0 \\ K = K_2|\gamma| + K_3; \tau\dot{\gamma} + \gamma = \mathrm{sign}(s); K_3 > 0; K_2 = K(t^*) & \text{if } = s = 0 \end{cases} \tag{6.4}$$

where t^* is the largest time value such that $s(t^{*-}) \neq 0$ and $s(t^*) = 0$.

Proof: See theorem 3 in [2].

6.3 Designing Adaptive SMC for Inverted Pendulum

Consider an inverted pendulum with the following structure:

$$\begin{aligned} \dot{x}_1 &= x_2 \\ \dot{x}_2\, g(x_1) &= u + \varphi_2 \tan x_1 - \varphi_3 x_2^2 \sin x_1 - dt \\ g(x_1) &= \varphi_1 \sec x_1 - \varphi_3 \cos x_1 \end{aligned} \tag{6.5}$$

To design the controller, the tracking error and Lyapunov function are defined as follows:

$$e = x_1 - \theta_d; s = ce + \dot{e}; \tag{6.6}$$

$$V = \frac{1}{2}g(x_1)s^2 + \frac{1}{2\gamma_1}(\varphi_1 - \hat{\varphi}_1)^2 + \frac{1}{2\gamma_2}(\varphi_2 - \hat{\varphi}_2)^2 + \frac{1}{2\gamma_3}(\varphi_3 - \hat{\varphi}_3)^2 \tag{6.7}$$

where $\gamma_i > 0$ and $\hat{\varphi}_i$ is the estimated value of φ_i. Choosing V_1 and V_2 as follows:

$$V_1 = \frac{1}{2}g(x_1)s^2; V_2 = \frac{1}{2\gamma_1}(\varphi_1 - \hat{\varphi}_1)^2 + \frac{1}{2\gamma_2}(\varphi_2 - \hat{\varphi}_2)^2 + \frac{1}{2\gamma_3}(\varphi_3 - \hat{\varphi}_3)^2 \tag{6.8}$$

The function derivatives are calculated as:

$$\dot{V}_1 = \frac{1}{2}\dot{g}(x_1)s^2 + g(x_1)s\dot{s}; \dot{V}_2 = -\frac{1}{\gamma_1}(\varphi_1 - \hat{\varphi}_1)\dot{\hat{\varphi}}_1 - \frac{1}{\gamma_2}(\varphi_2 - \hat{\varphi}_2)\dot{\hat{\varphi}}_2 - \frac{1}{\gamma_3}(\varphi_3 - \hat{\varphi}_3)\dot{\hat{\varphi}}_3 \tag{6.9}$$

Substituting $\dot{g}$ and $\dot{s}$ yields

$$\dot{V}_1 = \frac{1}{2}(\phi_1 \sec x_1 \tan x_1 + \phi_3 \sin x_1)x_2 s^2 + s\big(g(x_1)\big(c\dot{e} - \ddot{\theta}_d\big)$$
$$+ \big(u + \phi_2 \tan x_1 - \phi_3 x_2^2 \sin x_1 - dt\big)\big)$$
$$= \frac{1}{2}(\phi_1 \sec x_1 \tan x_1 + \phi_3 \sin x_1)x_2 s^2 + s\big((\phi_1 \sec x_1 - \phi_3 \cos x_1)\big(c\dot{e} - \ddot{\theta}_d\big)$$
$$+ \big(u + \phi_2 \tan x_1 - \phi_3 x_2^2 \sin x_1 - dt\big)\big)$$
$$= \phi_1\left(\frac{1}{2}x_2 s^2 \sec x_1 \tan x_1 + s \sec x_1 \big(c\dot{e} - \ddot{\theta}_d\big)\right) + \phi_2(s \tan x_1)$$
$$+ \phi_3\left(\frac{1}{2}x_2 s^2 \sin x_1 - s x_2^2 \sin x_1 - s \cos x_1 \big(c\dot{e} - \ddot{\theta}_d\big)\right) + s(u - dt) \quad (6.10)$$

u should be chosen in such a way that guarantees Lyapunov function's derivative to be negative. If we simplify the above structure as follows:

$$\dot{V}_1 = s\varphi_1 h_1 + \cdots + s\varphi_n h_n + s(u - dt) \quad (6.11)$$

Then a proper structure for u can be proposed as:

$$u = -\varphi \hat{1} h1 - \cdots - \hat{\varphi}_n h_n - \eta\, \mathrm{sgn}(s) \quad (6.12)$$

Consider a bounded disturbance $|dt| < \eta$. Substituting above controller results in:

$$\dot{V}_1 = s\tilde{\varphi}_1 h_1 + \cdots + s\tilde{\varphi}_n h_n - \eta s\, \mathrm{sgn}(s) - s dt \quad (6.13)$$

$$\dot{V}_2 = -\frac{1}{\gamma_1}\dot{\hat{\varphi}}_1 \tilde{\varphi}_1 - \frac{1}{\gamma_2}\dot{\hat{\varphi}}_2 \tilde{\varphi}_2 - \frac{1}{\gamma_3}\dot{\hat{\varphi}}_3 \tilde{\varphi}_3 \quad (6.14)$$

$$\dot{V}_1 + \dot{V}_2 = \tilde{\varphi}_1\left(sh_1 - \frac{1}{\gamma_1}\dot{\hat{\varphi}}_1\right) + \tilde{\varphi}_2\left(sh_2 - \frac{1}{\gamma_2}\dot{\hat{\varphi}}_2\right) + \tilde{\varphi}_3\left(sh_3 - \frac{1}{\gamma_3}\dot{\hat{\varphi}}_3\right)$$
$$- \eta s\, \mathrm{sgn}(s) - s dt \quad (6.15)$$

The parameter adaptation value is defined in the way that guarantees stability. To achieve negative Lyapunov derivative, the following adaptation laws are proposed:

$$\left.\begin{aligned} \dot{\hat{\varphi}}_1 &= \gamma_1 s h_1 \\ \dot{\hat{\varphi}}_2 &= \gamma_2 s h_2 \\ \dot{\hat{\varphi}}_3 &= \gamma_3 s h_3 \end{aligned}\right\} \dot{V}_1 + \dot{V}_2 = -\eta s \;\; \mathrm{sgn}(s) - s dt \overset{\eta > \max(dt)}{\Rightarrow} \dot{V}_1 + \dot{V}_2 < 0 \quad (6.16)$$

Figure 6.1 shows the block diagram of system. In Figure 6.2 the angular position tracking is illustrated using Adaptive SMC. As it is shown the system trajectory converges in 0.65 seconds. Figure 6.3 indicates the parameters adaptation. To illustrate the results more accurately we plot the first five seconds in Figure 6.4.

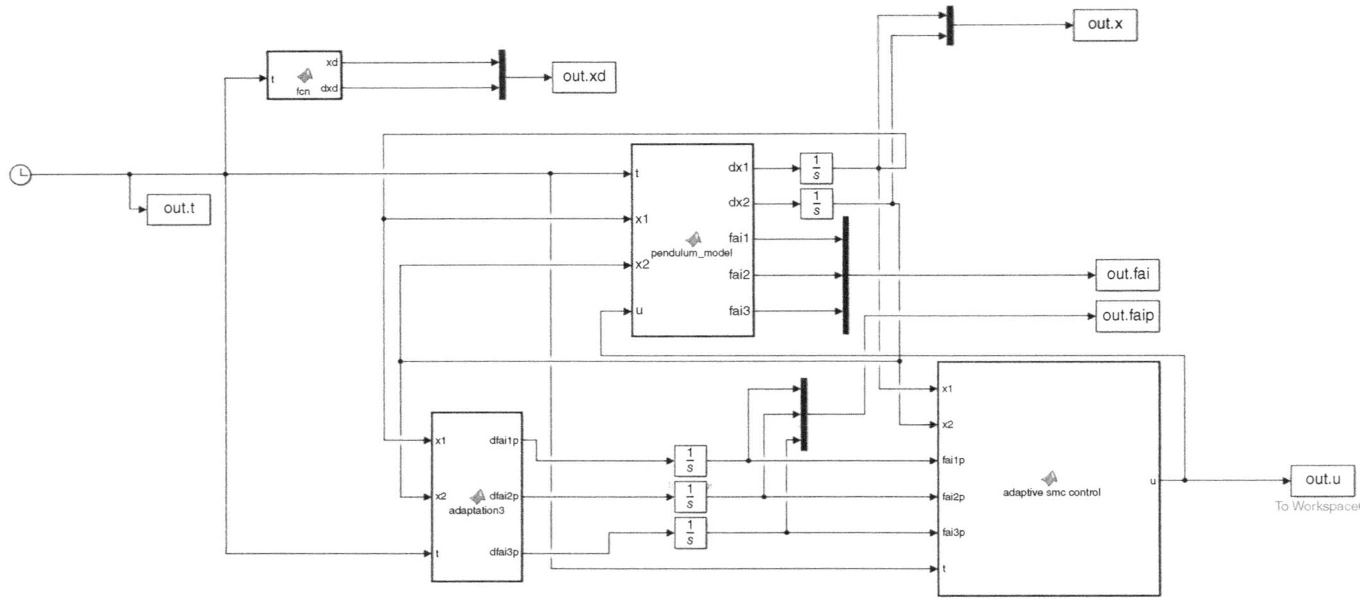

Figure 6.1 Systems block diagram in Simulink.

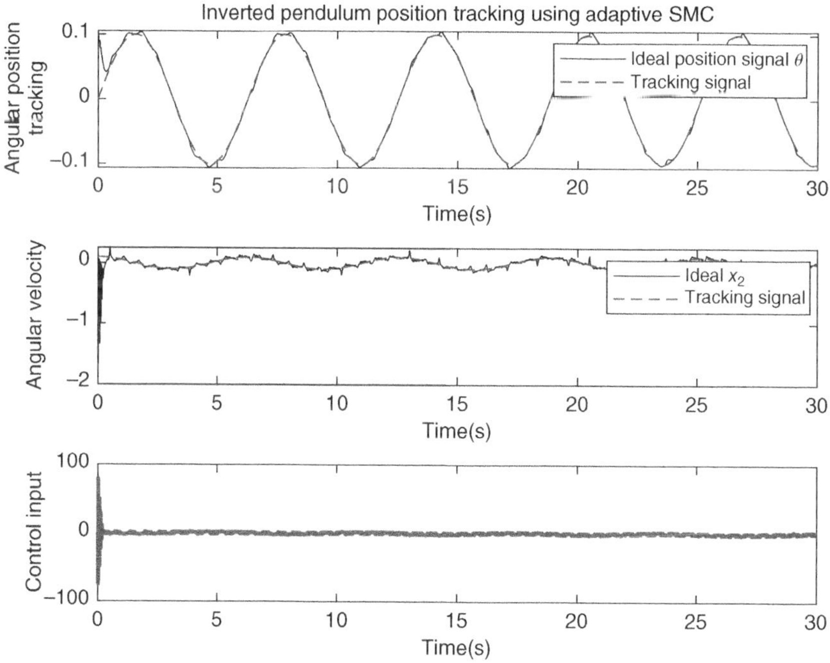

Figure 6.2 Inverted pendulum position tracking using adaptive SMC.

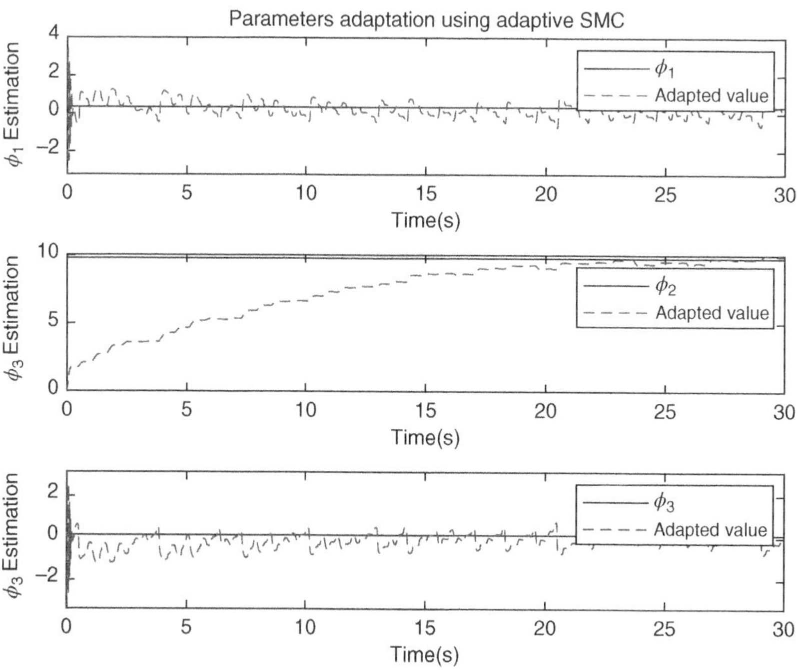

Figure 6.3 Parameter adaptation.

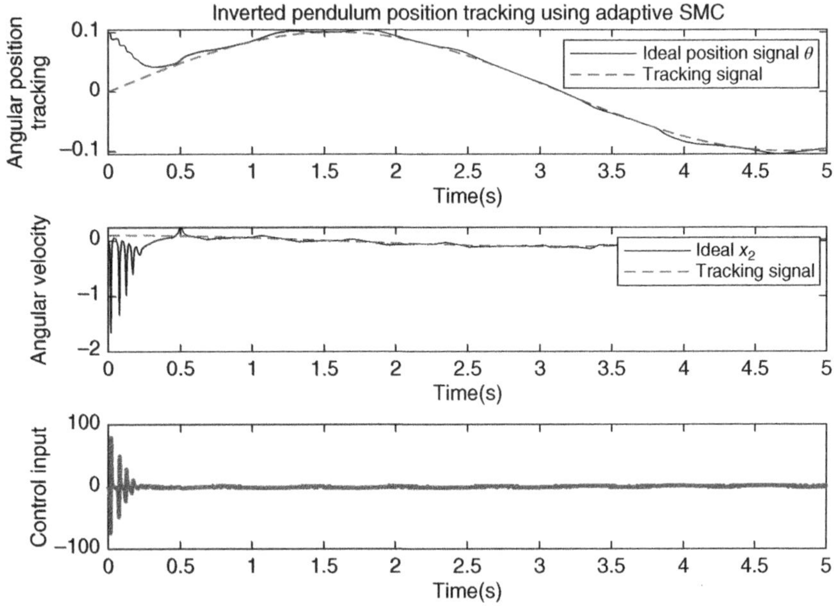

Figure 6.4 Inverted pendulum position tracking using adaptive SMC during first five minutes.

In the sequel, the related MATLAB functions are defined.

The function which is related to pendulum model is as follows:

```
function [dx1,dx2,fai1,fai2,fai3] = pendulum_model(u,t,
x1,x2)
mc = 0.5;mp = 0.5;
l = 0.3;
I = 0.006;
g = 9.8;
fai1 = (mc+mp)*(I+mp*l^2)/(mp*l);
fai2 = (mc+mp)*g;
fai3 = mp*l;

gx1 = fai1*sec(x1)-fai3*cos(x1);
dt = sin(10*t);

dx1 = x2;
dx2 = 1/gx1*(u-dt+fai2*tan(x1)-fai3*x2^2*sin(x1));
```

The adaptive sliding mode control function is defined as:

```
function u = adaptivesmccontrol(x1,x2,fai1p,fai2p,fai3p,t)
thd = 0.1*sin(t);
dthd = 0.1*cos(t);
ddthd = -0.1*sin(t);
e = x1-thd;
de = x2-dthd;
c = 10;
s = c*e+de;
xite = 1.0;
delta = 0.05;
kk = 1/delta;
u = -xite*sign(s)-fai1p*(0.5*s*x2*sec(x1)*tan(x1)+sec(x1)
*(c*de-ddthd))-fai2p*tan(x1)-fai3p*(0.5*s*x2*sin(x1)-
x2^2*sin(x1)-cos(x1)*(c*de-ddthd));
```

And parameters adaptation rules are considered as below:

```
function [dfai1p,dfai2p,dfai3p] = adaptation3(x1,x2,t)
thd = 0.1*sin(t);
dthd = 0.1*cos(t);
ddthd = -0.1*sin(t);
e = x1-thd;
de = x2-dthd;
c = 10;
s = c*e+de;
gama1 = 150;gama2 = 300;gama3 = 150;
dfai1p = gama1*(0.5*s^2*x2*sec(x1)*tan(x1)+s*sec(x1)*
(c*de-ddthd));
dfai2p = gama2*s*tan(x1);
dfai3p = gama3*(0.5*s^2*x2*sin(x1)-s*x2^2*sin(x1)-s*cos
(x1)*(c*de-ddthd));
```

Finally Simulation execution is done by the following MATLAB function:

```
clear, clc, close all
simout = sim('adaptivesmc3',[0 30]);

figure(1);
subplot(311);
plot(simout.t,simout.x(:,1),'k',simout.t,simout.xd
(:,1),'--b')
xlabel('time(s)');ylabel('angular Position tracking');
```

```
legend('Ideal position signal \theta','tracking signal');
title('Inverted Pendulum position Tracking using adaptive
SMC');
subplot(312);
plot(simout.t,simout.x(:,2),'k',simout.t,simout.xd
(:,2),'--b')
xlabel('time(s)');ylabel('angular velocity');
legend('Ideal x2','tracking signal');
subplot(313)
plot(simout.t,simout.u,'linewidth',2);
xlabel('time(s)');ylabel('control input');


figure(2);

subplot(311);
plot(simout.t,simout.fai(:,1),'k',simout.t,simout.faip
(:,1),'--b')
xlabel('time(s)');ylabel('\phi_1 estimation');
legend('\phi_1','adapted value');
title('Parameters adaptation using adaptive SMC');
subplot(312);
plot(simout.t,simout.fai(:,2),'k',simout.t,simout.faip
(:,2),'--b')
xlabel('time(s)');ylabel('\phi_3 estimation');
legend('\phi_2','adapted value');
subplot(313);
plot(simout.t,simout.fai(:,3),'k',simout.t,simout.faip
(:,3),'--b')
xlabel('time(s)');ylabel('\phi_3 estimation');
legend('\phi_3','adapted value');
```

References

1 Utkin (2022). *Road Map for Sliding Mode Control Design*. Springer Link.
2 Plestan, F., Shtessel, Y., Bregeault, V., and Poznyak, A. (2010). New methodologies for adaptive sliding mode control. *International Journal of Control* 83 (9): 1907–1919.
3 Utkin, V.I. and Poznyak, A.S. (2013). Advances in sliding mode control: concept, theory and implementation. In: *Adaptive Sliding Mode Control*. Berlin: Springer.

7

Super-Twisting Sliding Mode Controller

7.1 Introduction

In this chapter, the super-twisting sliding mode control technique is introduced and applied on an inverted pendulum. In recent years, many researchers have studied various control methods on highly nonlinear and unstable systems, including the inverted pendulum. Design of an efficient controller with suitable performance requires calculations, the designer's precise understanding of the controller, and examination of the issues. In this chapter a super-twisting sliding mode controller (STSMC) is defined, and the performance and effectiveness of the method are discussed in nonlinear interference interpretation, error reduction, and control adjustment fields. Sliding mode controller is a robust method to deal with disturbances and uncertainties with a known range of the system [1]. One of the disadvantages of this controller is the chattering phenomenon, which reduces the resistance of the controller. In many researches such as [2], in order to deal with disturbances, uncertainties, and also chattering phenomenon as the main factor in increasing energy consumption and reducing the speed of convergence of the system, the super-twisting sliding mode controlling method is used. The super-twisting algorithm was first introduced in 1986 by Emelyanov et al. and later developed by Levantovsky [3] from the same team. The results of applying the proposed controller on the inverted pendulum have been investigated in terms of tracking speed, convergence time, error, and chattering reduction. In addition, the stability analysis of the closed-loop system is presented according to the Lyapunov theorem. In this chapter, in Section 7.2, the SMC methodology is reviewed [4, 5] is fully introduced. In addition, in this section, the design of controller and the proof of the stability of the controller are done. In Section 7.3, the equations related to the STSMC method are explained. Finally, in Section 7.4, the results are simulated and evaluated.

Methods of Developing Sliding Mode Controllers: Design and MATLAB Simulation, First Edition.
Reihaneh Kardehi Moghaddam and Mostafa Rabbani.

Published 2025 by John Wiley & Sons, Inc.

7.2 Sliding Mode Controller Review

7.2.1 Sliding Mode Controller Design

For designing SMC, we follow Chapter 2. As it was defined previously, the following theorem can be considered for controller design.

Theorem 7.1
For canonical nonlinear system (7.1), the proposed controlling structure (7.2) is a stabilizer controller.

$$\begin{aligned} \dot{x}_1 &= x_2 \\ \dot{x}_2 &= f + gu + d \end{aligned} \tag{7.1}$$

$$u = \hat{u} - \frac{k}{\hat{g}}\text{sign}(s) \Rightarrow u = \frac{1}{\hat{g}}\left(-\hat{f} + \ddot{x}_d - \lambda\dot{\tilde{x}} - k\,\text{sign}(s)\right), k \geq F + \eta + D \tag{7.2}$$

Proof: To prove the stability consider the Lyapunov function

$$V = \frac{1}{2}s^2(t)$$

Then it should be proved that u, which is defined in (7.2), will lead to a negative derivative of the Lyapunov function

$$\dot{V} = \frac{d}{dt}\left(\frac{1}{2}s^2(t)\right) = s(t)\dot{s}(t) \leq -\eta|s(t)|$$

$$\Rightarrow s\left(f + gu + d - \ddot{x}_d + \lambda\dot{\tilde{x}}\right) \leq -\eta|s| \rightarrow$$

$$\Rightarrow s\left(f + g\left(\hat{u} - \frac{k}{\hat{g}}\text{sign}(s)\right) + d - \ddot{x}_d + \lambda\dot{\tilde{x}}\right) \leq -\eta|s| \rightarrow$$

$$\Rightarrow s\left(f + g\left(\frac{1}{\hat{g}}\left(-\hat{f} + \ddot{x}_d - \lambda\dot{\tilde{x}}\right) - \frac{k}{\hat{g}}\text{sign}(s)\right) + d - \ddot{x}_d + \lambda\dot{\tilde{x}}\right) \leq -\eta|s|$$

$$\Rightarrow s\left(f - g\hat{g}^{-1}\hat{f} + g\hat{g}^{-1}\ddot{x}_d - \lambda g\hat{g}^{-1}\dot{\tilde{x}} - kg\hat{g}^{-1}\text{sign}(s) + d - \ddot{x}_d + \lambda\dot{\tilde{x}}\right) \leq -\eta|s|$$

$$\Rightarrow s\left(f - g\hat{g}^{-1}\hat{f} - kg\hat{g}^{-1}\text{sign}(s) + d + (g\hat{g}^{-1} - 1)\ddot{x}_d - (g\hat{g}^{-1} - 1)\lambda\dot{\tilde{x}}\right) \leq -\eta|s|$$

Adding and subtracting $\hat{f}$ yield:

$$\Rightarrow s\left(f - \hat{f} + (g\hat{g}^{-1} - 1)\underbrace{\left(-\hat{f} + \ddot{x}_d - \lambda\dot{\tilde{x}}\right)}_{\hat{u}} - kg\hat{g}^{-1}\,\text{sign}(s) + d\right) \leq -\eta|s|$$

$$\Rightarrow s\left(f-\hat{f}+(g\hat{g}^{-1}-1)\underbrace{\left(-\hat{f}+\ddot{x}_d-\lambda\dot{\tilde{x}}\right)}_{\hat{u}}-kg\hat{g}^{-1}\operatorname{sign}(s)+d\right)\leq-\eta|s|$$

$$\Rightarrow s\left(f-\hat{f}+(g\hat{g}^{-1}-1)\hat{u}\right)-ksg\hat{g}^{-1}\operatorname{sign}(s)+sd\leq-\eta|s|$$

$$\Rightarrow |s|\left|f-\hat{f}+(g\hat{g}^{-1}-1)\hat{u}\right|-kg\hat{g}^{-1}|s|+|d||s|\leq-\eta|s|\rightarrow$$

Using triangular inequality we have:

$$\Rightarrow |s|\underbrace{\left|f-\hat{f}\right|}_{F}+\left|g\hat{g}^{-1}-1\right||\hat{u}||s|-kg\hat{g}^{-1}|s|+D|s|\leq-\eta|s|$$

$$\Rightarrow\begin{cases} s>0\xrightarrow{s=|s|}F+D+\left|g\hat{g}^{-1}-1\right||\hat{u}|\leq k-\eta \\ s<0\xrightarrow{s=-|s|}-F-D-\left|g\hat{g}^{-1}-1\right||\hat{u}|\geq k-\eta \end{cases}$$

$$\Rightarrow k\geq\beta(F+\eta+D)+(\beta-1)|\hat{u}|g$$

$$\Rightarrow \forall\beta=1, k\geq F+\eta+D$$

So the proof is done and the proposed controller leads to $\dot{V}<0$ and increase of V.

7.2.2 Inverted Pendulum Design

The model discussed in this chapter is a nonlinear unstable inverted pendulum system. In this section, the dynamic equations of the inverted pendulum system are introduced and the nonlinear behavior of the state variables and their dependence on the initial conditions are investigated.

Figure 7.1 shows an inverted pendulum structure.

The equations of the inverted pendulum system are in the form of (7.3).

$$\begin{cases} \dot{x}_1=x_2 \\ \dot{x}_2=f(x_1,x_2)+b(x_1,x_2)U \end{cases} \qquad \dot{x}=\begin{bmatrix}\theta\\ \dot{\theta}\end{bmatrix}$$

$$\begin{cases} f(x_1,x_2)=\dfrac{(M+m)g\sin x_1-mLx_2^2(\sin x_1\cos x_1)}{\frac{4}{3}(M+m)L-mL\cos^2 x_1} \\ b(x_1,x_2)=\dfrac{\cos x_1}{\frac{4}{3}(M+m)L-mL\cos^2 x_1} \end{cases} \tag{7.3}$$

In above relations, functions f and b are nonlinear equations expressed in the form of Eq. (7.3). As can be seen in Figure 7.2, for different starting points, the states converge to an oscillatory limit cycle.

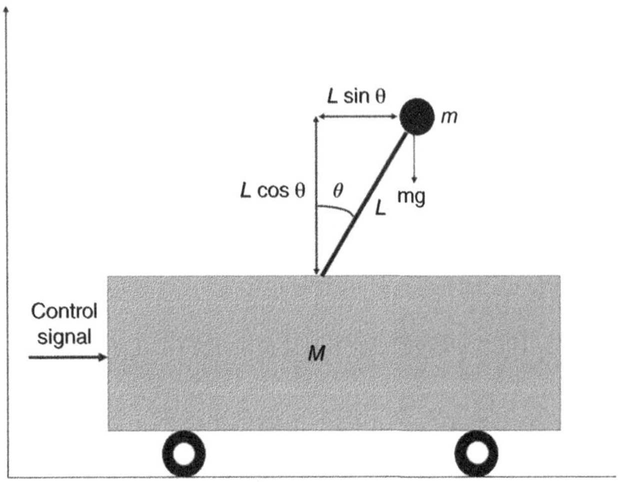

Figure 7.1 Physical model of inverted pendulum.

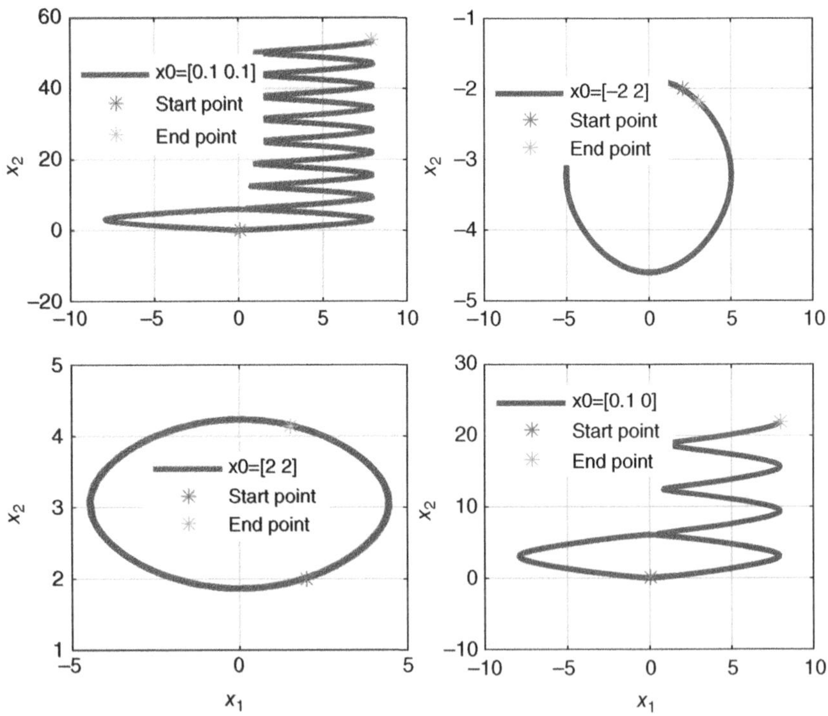

Figure 7.2 Dynamic limit cycle of the nonlinear inverted pendulum system for different initial conditions.

Here, the goal is to control the angle of the pendulum so that the pendulum follows the behavior of the input reference signal and the tracking error converges to zero. In the next part a super-twisting controller is designed for inverted pendulum, and the results are compared with SMC.

7.3 Super-Twisting Sliding Mode Controller

Due to the presence of the sign function in the control input, the occurrence of the undesirable phenomenon of chattering in the control input of the inverted pendulum system is inevitable. Therefore, in the following, the super-twisting algorithm is presented, which eliminates the chattering phenomenon in the control input and reduces its range [6, 7].

The most important advantage of the super-twisting sliding mode compared to the classic sliding mode is the elimination of the chattering phenomenon. Using sign function in SMC design causes chattering phenomenon. Chapter 9 will fully discuss chattering removal. The super-twisting algorithm is presented for control systems with relative degree one in order to reduce the chattering problem. In this method, the system-state trajectory rotates around the origin, and in each revolution, the absolute value of the intersection distance of the paths with the axes decreases and finally converges to the origin according to Figure 7.3.

Control law and the equations related to the super-twisting algorithm are as follows [8–10]:

$$u_{ST} = -\alpha|s|^{P}\,\text{sign}(s) + u_1 \tag{7.4}$$

Figure 7.3 Super-twisting algorithm phase trajectory.

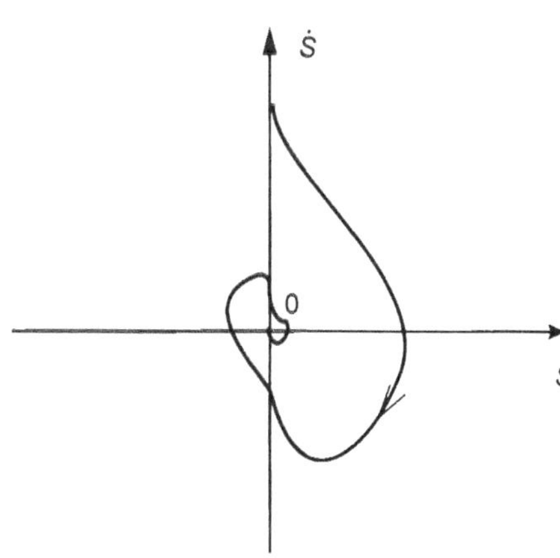

$$\dot{u}_1 = -\omega\,\text{sign}(s) \tag{7.5}$$

$$u_{ST} = -\alpha|s|^P\,\text{sign}(s) - \omega\int \text{sign}(s)d\tau \tag{7.6}$$

$$u = \hat{u} - u_{ST} \tag{7.7}$$

Theorem 7.2
Consider system (7.1). The proposed controlling structure (7.6) will stabilize system (7.1).

Proof: To prove the stability the controller (7.6) reachability condition $s(t)\dot{s}(t) \le -\mu|s(t)|$ should be satisfied:

$$s(\lambda(\dot{x}_1 - \dot{x}_d) + (\dot{x}_2 - \ddot{x}_d)) \le -\mu|s| \tag{7.8}$$

$$s(\lambda(x_2 - \dot{x}_d) + (f + bu - \ddot{x}_d)) \le -\mu|s| \tag{7.9}$$

$$s\left(\lambda(x_2 - \dot{x}_d) + \left(f + b\left(\frac{-1}{b}(\lambda(x_2 - \dot{x}_d) + (f - \ddot{x}_d)\right) - \alpha|s|^\rho\text{sign}(s) - \omega\int \text{sign}(s)d\tau\right) - \ddot{x}_d\right) \tag{7.10}$$

$$s\left[\lambda x_2 - \lambda\dot{x}_d + f - \lambda x_2 + \lambda\dot{x}_d - f + \ddot{x}_d + \alpha|s|^\rho\text{sign}(s) + \omega\int \text{sign}(s)d\tau - \ddot{x}_d\right] \le -\mu|s| \tag{7.11}$$

$$s\left[\alpha|s|^\rho\text{sign}(s) + \omega\int \text{sign}(s)d\tau\right] \le -\mu|s| \tag{7.12}$$

$$[\alpha|s|^\rho|s| + \omega|s|] \le -\mu|s| \tag{7.13}$$

$$[\alpha|s|^\rho + \omega] \le -\mu \tag{7.14}$$

The parameters are set as follows. By choosing appropriate parameters in Eq. (7.3), reachability inequality is confirmed. By choosing the following parameters, the speed of convergence is optimized. Also, the global asymptotic stability of system (7.1) to the controller (7.6) is guaranteed, and the tracking error is bounded. □

How the state variables move in the phase plane $(S, \dot{S})$ for the super-twisting algorithm is shown in Figure 7.3. In the next section, the proposed controller is simulated using MATLAB.

7.4 Super-Twisting SMC Design in MATLAB

In this section, the controller designed in Section 7.3 is tested on the inverted pendulum system. Accordingly we suppose $x_0 = [0.2, 0.1]$ and $x_d = \frac{\pi}{30}\sin(t)$. X_0 is the initial condition and X_d is the input reference signal.

7.4.1 MATLAB Programming

The simulation of the above system by the Simulink section in MATLAB software is shown in Figure 7.4.

The input signal function program is as follows.

```
function [wdot,u,s] = fcn(x1,x2,e,de,dyd,ddyd,w)

m = 0.1; M = 1; L = 0.5; g = 9.8; k = 18; y = 5;

%lambda s = de+y*e;

%sliding surface ust = -2*(abs(s))^0.5*sign(s)+w;

%Super twisting control wdot  = -8*sign(s);

f = ((M+m)*g*sin(x1)-m*L*(x2^2)*sin(x1)*cos(x1))/((4/3)*
(M+m)*L-m*L*(0.5+0.5*cos(2*x1)));

b = cos(x1)/((4/3)*(M+m)*L-m*L*(0.5+0.5*cos(2*x1)));
u = (1/b)*(-f+ddyd-y*de+ust)
```

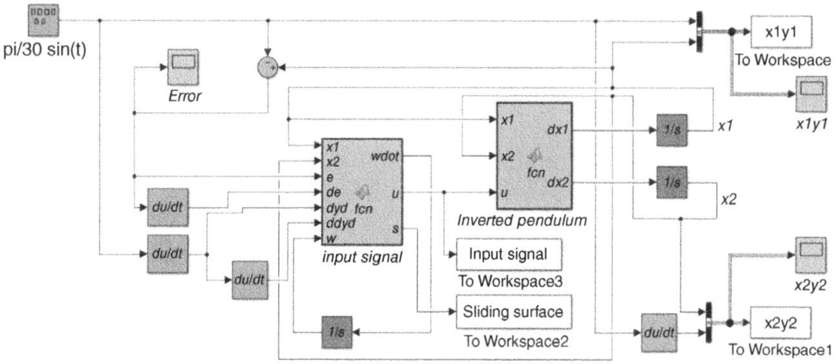

Figure 7.4 Block diagram of the system in Simulink.

The inverted pendulum function program is as follows.

```
function [dx1,dx2]= fcn(x1,x2,u)
m = 0.1;
M = 1;
L = 0.5;
g = 9.8;
zeros(2,1);
dx1 = x2;
f =((M+m)*g*sin(x1)-m*L*(x2^2)*sin(x1)*cos(x1))/((4/3)*
(M+m)*L-m*L*(0.5+0.5*cos(2*x1)));
b = cos(x1)/((4/3)*(M+m)*L-m*L*(0.5+0.5*cos(2*x1)));
dx2 = f+b*u;
```

In the following, the sliding surface and the control signal of the inverted pendulum are discussed using the method mentioned in the above equations.

Figure 7.5 shows that STSMC contains less chattering in specified points. Additionally Figure 7.6 shows acceptable convergence of sliding surface to zero using the proposed method.

The existence of the sign function causes very large fluctuations in the input control signal, and these fluctuations are very harmful for control systems in industry. By removing the chattering phenomenon, the tracking error is improved according to Figure 7.7, and the time to reach the sliding surface is reduced, which makes the controller more resistant to uncertainties and disturbances.

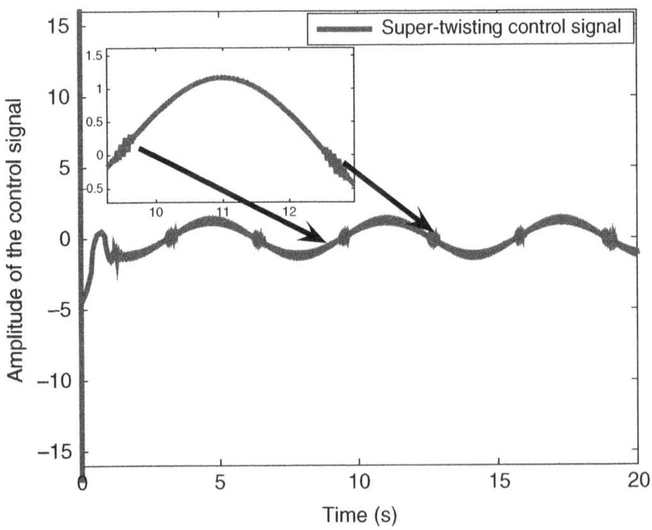

Figure 7.5 Control signal in super-twisting SMC.

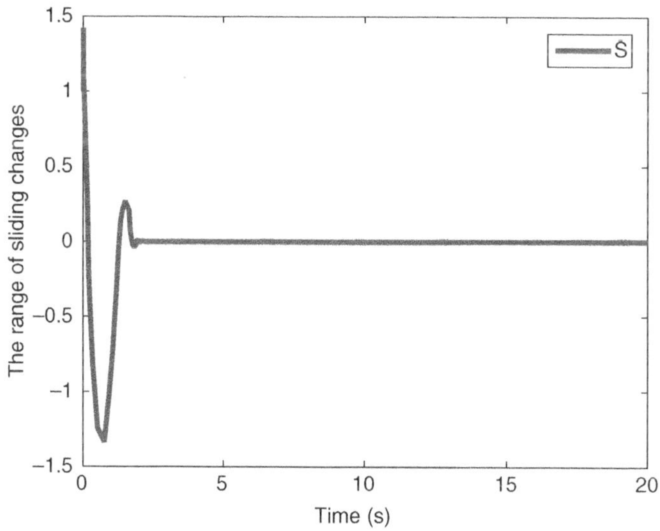

Figure 7.6 Sliding surface in super-twisting SMC.

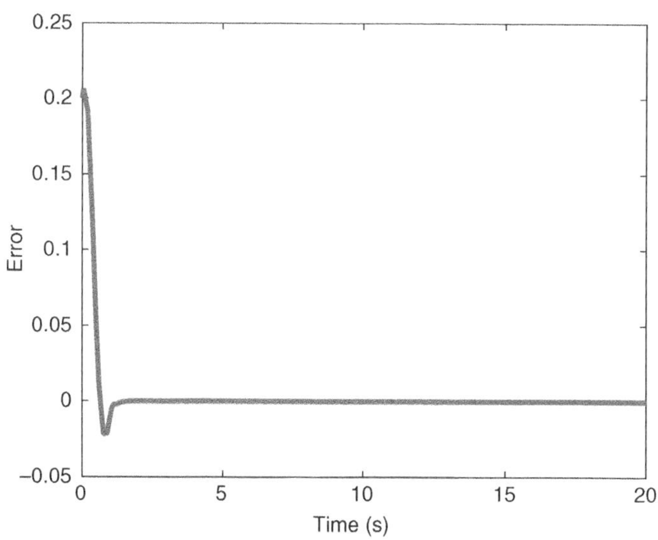

Figure 7.7 The tracking error.

As shown in Figure 7.7, the tracking error of the input reference signal converges to zero in less than 2 seconds.

Figures 7.8 and 7.9 show the performance of the super-twisting control method with the lowest amount of error and the best accuracy, which completely and with negligible error follows the desired input.

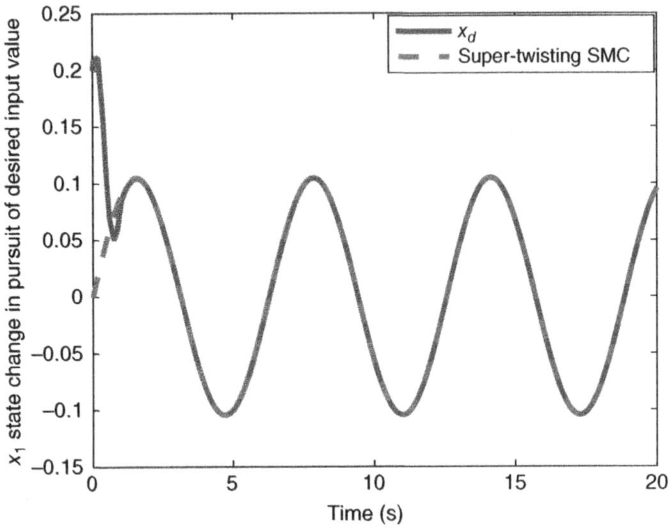

Figure 7.8 The result of tracking the reference signal by x_1.

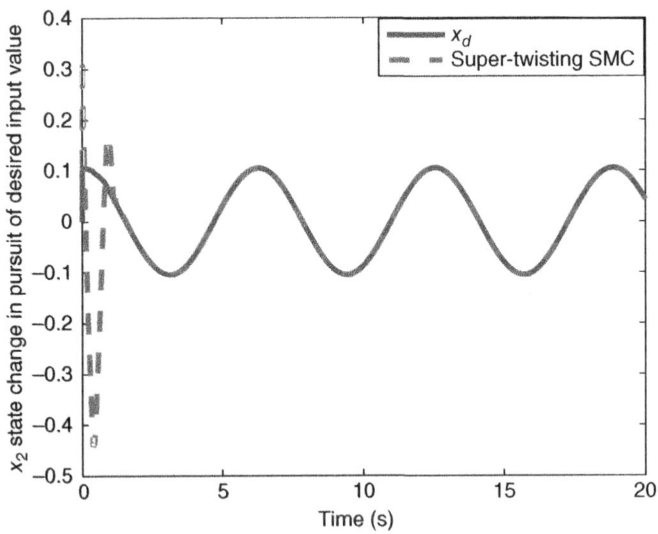

Figure 7.9 The result of tracking the reference signal by x_2.

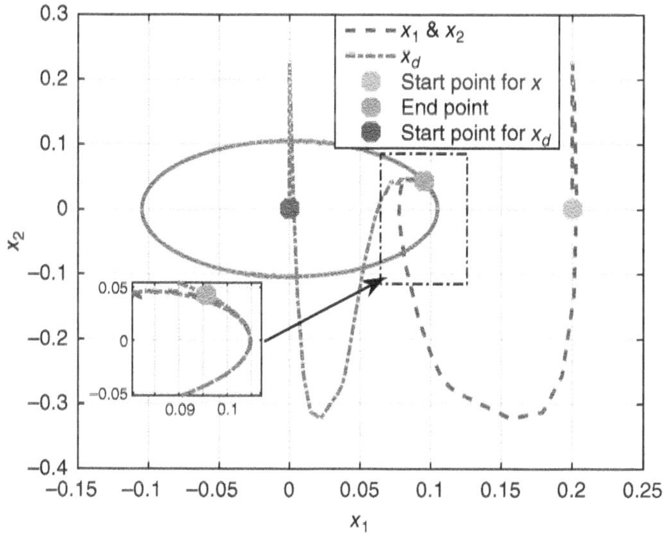

Figure 7.10 2D diagram of reference tracking.

In Figure 7.10, this tracking operation is drawn in 2D and the time to reach the desired value of the input reference signal is specified. In Table 7.1, the convergence time to the sliding surface with an error less than $e = 0.0001$ has been examined.

According to Table 7.1, the state variables in the super-twisting method reach the sliding surface in less than 2 seconds and are attracted to this surface. Table 7.2 shows the performance analysis of the control method in terms of steady-state error, reaching time, control signal overshoot, controller peak-to-peak range, and MSE error, where N shows the number of sampled state values.

$$\text{MSE} = \frac{1}{N}\sum_{i=1}^{n}(X - X_d)^2 \tag{7.15}$$

According to Table 7.2, in terms of accuracy and convergence speed, the STSMC method shows a very good performance. Also, the overshoot values and changes in the control signal range are acceptable in this method, and it effectively controls the system and eliminates the chattering phenomenon. Although there is a chattering phenomenon.

Table 7.1 Analysis of the convergence speed.

Response specification	Super-Twisting SMC
Convergence time to sliding surface	1.35 s

Table 7.2 Performance analysis of control strategies.

Response specification	**Steady-state error**	**Reaching time (sec)**	**Over shoot(u)**	**Peak-to-peak amplitude of the u signal**	**MSE**
Super-twisting SMC	**0**	**1.67**	**1.176**	**2.351**	0.0096

References

1 Shamsabadi, M. and Moghaddam, R.K. (2021). Underwater robot trajectory control using non-singular terminal sliding mode controller. *International Journal of Smart Electrical Engineering* 10 (3): 117–125.

2 Chaal, H. and Jovanovic, M (2010). Second order sliding mode control of a DC drive with uncertain parameters and load conditions. *2010 Chinese Control and Decision Conference.*

3 Levant, A. (1993). Sliding order and sliding accuracy in sliding mode control. *International Journal of Control* 58 (6): 1247–1263.

4 Shahri, A.P., Haghighatnia, S., Moghaddam, R.K., and Kobravi, H.R. (2017). Control the tumor growth via sliding mode control. *International Journal of Medical Engineering and Informatics* 9 (2).

5 Razzaghian, A., Moghaddam, R.K., and Pariz, N. (2022). Adaptive fuzzy fractional-order fast terminal sliding mode control for a class of uncertain nonlinear systems. *International Journal of Industrial Electronics, Control and Optimization* 5 (1): 77–87.

6 Mingyuan, H., Ahn, H., Chung, Y., and You, K. (2023). Speed regulation for PMSM with super-twisting sliding-mode controller via disturbance observer. *Mathematics* 11 (7): 1618.

7 Liu, W., Ye, H., and Yang, X. (2023). Super-twisting sliding mode control for the trajectory tracking of underactuated USVs with disturbances. *Journal of Marine Science and Engineering* 11: 636.

8 Fei, J. and Feng, Z. (2019). *Adaptive Fuzzy Super-Twisting Sliding Mode Control for Microgyroscope*. Wiley, Hindawi.

9 Manzanilla, A., Ibarra, E., Salazar, S. et al. (2021). Super-twisting integral sliding mode control for trajectory tracking of an unmanned underwater vehicle. *Ocean Engineering* 234: 109164.

10 Boiko, I., Fridman, L., Pisano, A., and Usai, E. (2007). Analysis of chattering in systems with second-order sliding modes. *IEEE Transactions on Automatic Control* 52 (11): 2085–2102.

8

Back-Stepping Sliding Mode Control

8.1 Introduction

Since the kinematic and dynamic equations of most industrial systems are nonlinear in nature, using nonlinear control methods is unavoidable. In practice there are various disturbances and uncertainties in the system. Therefore it is essential for the designed controllers to withstand these uncertainties and disturbances. One such controller that is resistant to uncertainties and disturbances and does not exhibit chattering behavior is back-stepping controller. In control science, back-stepping is a robust method designed in 1990 by Petar V. Kokotovic and his colleagues for a special class of nonlinear systems [1–2]. In this method, the main system is divided into several subsystems and stabilized using Lyapunov-based designing strategies. In this recursive structure, the designer can start the design process with a known stable system and recursively develop new controllers. Generally, in this method, the controller is designed in the final stage. Due to this approach, this control is called back-stepping [3]. One of the most significant features of the back-stepping method is that it avoids eliminating nonlinear terms, which sets it apart from the feedback linearization method. Eliminating nonlinear terms in feedback linearization schemes is common and requires a significant amount of control effort. The resulting back-stepping controllers require less control effort than linear feedback controllers. The high efficiency of this method can be attributed to its simple design, step-by-step implementation, and the advantage it offers in reducing chattering. The back-stepping method is a systematic Lyapunov-based technique for designing stable controls of nonlinear dynamic systems. Back-stepping control technique is widely used to achieve the goals of industrial nonlinear system control including military equipment, robotics, medical engineering, and technical services [4, 5]. In this chapter, Section 8.1 introduces the back-stepping super-twisting SMC method, its applications, and history. In

Methods of Developing Sliding Mode Controllers: Design and MATLAB Simulation, First Edition.
Reihaneh Kardehi Moghaddam and Mostafa Rabbani.

Published 2025 by John Wiley & Sons, Inc.

Section 8.2, the back-stepping method is fully introduced, and the dynamic equations of the chaotic Duffing oscillator are described and simulated. In addition, in Section 8.3, the design of the controller of the chaotic Duffing system and the proof of stability are discussed. In Section 8.4, the controller designed for chaotic Duffing system is simulated. Finally, in Section 8.5, with the help of back-stepping super-twisting SMC method, the controller is designed and simulated for the inverted pendulum, and the results are compared with feedback linearization in terms of steady-state error, control signal amplitude, MSE error, and convergence time.

8.2 Back-Stepping Controller Design

In Section 8.2.1, the back-stepping method is fully described and then, in Section 8.2.2, the equations of the chaotic Duffing oscillator [6] are described and simulated with MATLAB software.

8.2.1 Back-Stepping Method

Back-stepping is a recursive controller design technique. Suppose the structure of the system as follows [3, 7]:

$$\begin{aligned} \dot{\eta} &= f(\eta) + g(\eta)\zeta \\ \dot{\zeta} &= u \end{aligned} \tag{8.1}$$

Here $[\eta, \zeta]^T$ are the system states and u is the input control signal. The functions f and g are smoothed and known. Suppose there is a state feedback control as (8.2) such that system (8.3) is asymptotically stable.

$$\zeta = \phi(\eta) \tag{8.2}$$

$$\dot{\eta} = f(\eta) + g(\eta)\phi(\eta) \tag{8.3}$$

A positive definite Lyapunov function is chosen to apply to the following equation.

$$\dot{V}_1(\eta) = \frac{\partial V}{\partial \eta}[f(\eta) + g(\eta)\phi(\eta)] \leq -W(\eta) \tag{8.4}$$

The function $W(\eta)$ is positive definite. By adding and subtracting term $g(\eta)\phi(\eta)$ to Eq. (8.1), (8.1) converts to (8.5).

$$\begin{aligned} \dot{\eta} &= [f(\eta) + g(\eta)\phi(\eta)] + g(\eta)[\zeta - \phi(\eta)] \\ \dot{\zeta} &= u \end{aligned} \tag{8.5}$$

Here, a variable named z is defined as follows:

$$z = \zeta - \phi(\eta), \quad \dot{z} = \dot{\zeta} - \dot{\phi}(\eta) = u - \dot{\phi} \tag{8.6}$$

As a result, system (8.5) is rewritten as follows:

$$\begin{aligned} \dot{\eta} &= [f(\eta) + g(\eta)\phi(\eta)] + g(\eta)z \\ \dot{z} &= u - \dot{\phi} \end{aligned} \tag{8.7}$$

Here f, g, ϕ are known and the derivative ϕ is calculated as follows:

$$\dot{\phi} = \frac{\partial \phi}{\partial \eta}[f(\eta) + g(\eta)\zeta] \tag{8.8}$$

It is enough to put a virtual controller ν instead of $u - \dot{\phi}$ to make a cascade structure. The virtual controller is defined as follows:

$$\nu = \dot{z} \tag{8.9}$$

System (8.7) is rewritten as follows:

$$\begin{aligned} \dot{\eta} &= [f(\eta) + g(\eta)\phi(\eta)] + g(\eta)z \\ \dot{z} &= \nu \end{aligned} \tag{8.10}$$

Consider the below Lyapunov function to stabilize the system (8.10).

$$V_a(\eta, \zeta) = V(\eta) + \frac{1}{2}z^2 \tag{8.11}$$

If the derivative of Eq. (8.11) becomes negative, it is guaranteed that z is decreasing and the system will be stable.

$$\dot{V}_a = \frac{\partial V}{\partial \eta}[f(\eta) + g(\eta)\phi(\eta)] + \frac{\partial V}{\partial \eta}g(\eta)Z + Z\nu \leq -W(\eta) + \frac{\partial V}{\partial \eta}g(\eta)Z + Z\nu \tag{8.12}$$

The virtual controller v is chosen in a manner that always ensures Eq. (8.12) is negative. Therefore, the virtual controller is defined as follows:

$$\nu = -\frac{\partial V}{\partial \eta}g(\eta) - KZ, K > 0 \tag{8.13}$$

Eventually

$$\dot{V}_a \leq -W(\eta) - KZ^2 \tag{8.14}$$

Equation (8.14) is always negative.

By setting ν, z, and ϕ, the feedback control law is calculated as follows:

$$u = \frac{\partial \phi}{\partial \eta}[f(\eta) + g(\eta)\zeta] - \frac{\partial V}{\partial \eta}g(\eta) - K[\zeta - \phi(\eta)] \tag{8.15}$$

8.2.2 Chaotic Duffing Oscillator Equations

The Duffing equation (or Duffing oscillator), named after Georg Duffing (1861–1944), is a nonlinear second-order differential equation used to model certain damped and driven oscillators. Consider a second-order chaotic system defined by Duffing equation [8, 9]:

$$\begin{aligned} \dot{x}_1 &= x_2 \\ \dot{x}_2 &= -p_1 x_2 - p_2 x_1 - p_3 x_1^3 + p_4 \cos(\omega t) + u \end{aligned} \tag{8.16}$$

Equation (8.16) is equivalent as follows:

$$\begin{aligned} \dot{x}_1 &= x_2 \\ \dot{x}_2 &= f(x_1, x_2) + g(x_1, x_2)u \end{aligned} \tag{8.17}$$

Equation (8.16) for values $p_1 = 0.4, p_2 = -1.1, p_3 = 1, p_4 = 2.1, \omega = 1.8, u = 0$ has a chaotic behavior. The chaotic behavior of the Duffing system and its limit cycle are shown in Figures 8.1 and 8.2. The system is simulated for the initial condition $X_0 = [0.2, 0]$.

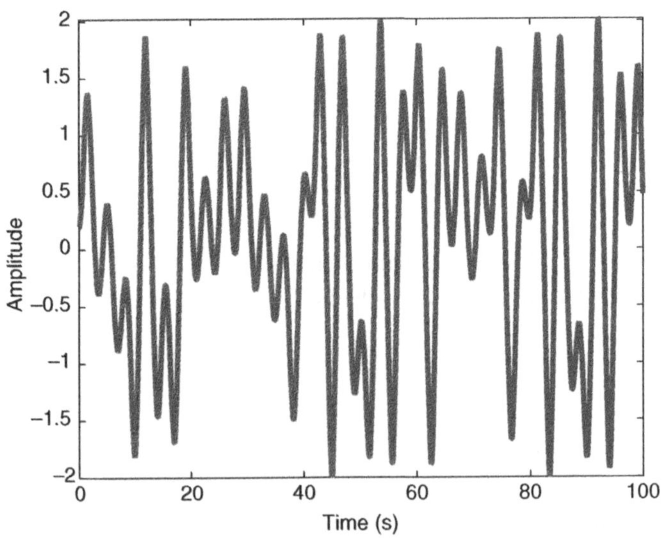

Figure 8.1 Chaotic behavior of the Duffing system.

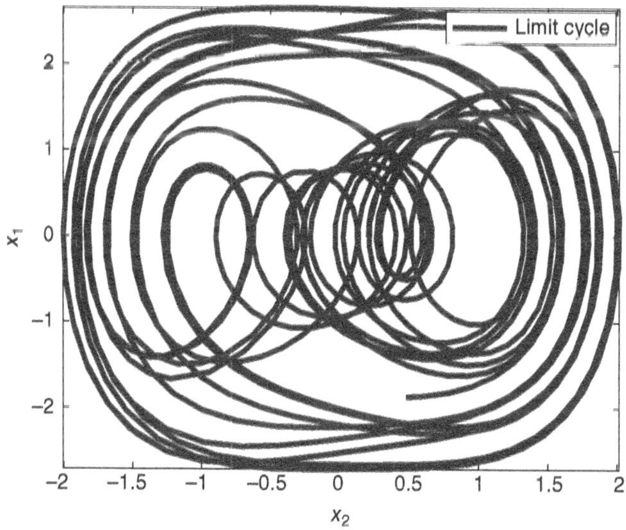

Figure 8.2 Limit cycle of the state variables of the Duffing system.

The MATLAB program for drawing the above diagram is as follows. The following program describes the equations of the chaotic Duffing system:

```
function dx = duffing(t,x)
global p1 p2 p3 p4 w
dx = zeros(2,1);
dx(1) = x(2);
dx(2) = -p1*x(2)-p2*x(1)-p3*x(1)^3+p4*cos(w*t);
end
```

The following program draws related graphs:

```
clc
clear all
close all

global p1 p2 p3 p4 w

p1 = 0.4;
p2 = -1.1;
p3 = 1;
p4 = 2.1;
w = 1.8;
```

```
tspan = [0 100];
init = [0.2 0];

[t x] =ode45(@duffing,tspan,init);

figure(1)
plot(t,x(:,1),'r')

axis tight
title('time series')
xlabel('time(sec)')
ylabel('x')

figure(2)

plot(x(:,1),x(:,2),'b')

axis tight
title('phase space')
xlabel('x2')
ylabel('x1')
```

8.3 Back-Stepping Control of Chaotic Duffing Oscillator

In this section, the back-stepping controller is designed for the chaotic Duffing oscillator, and the stability of the designed controller is proved. Here, the control objective is to track a desired input signal such that the tracking error converges to zero [10]. Therefore, in the first step, a tracking error must be introduced according to Eq. (8.18).

$$z_1 = x_1 - x_d \tag{8.18}$$

The error dynamics is described as the following equation.

$$\dot{z}_1 = \dot{x}_1 - \dot{x}_d = x_2 - \dot{x}_d \tag{8.19}$$

The expression x_2 is the virtual controller for which the stabilizer function should be designed. The stabilizer function is introduced as follows. In fact, α is the desired value of x_2.

$$\alpha = \dot{x}_d - k_1 z_1 \tag{8.20}$$

The difference between the actual value and the desired value is defined as the second error variable in the form of the following equation:

$$z_2 = x_2 - \alpha = x_2 + k_1 z_1 - \dot{x}_d \tag{8.21}$$

So

$$\dot{z}_1 = \dot{x}_1 - \dot{x}_d = \alpha - \dot{x}_d + z_2 = z_2 - k_1 z_1 \tag{8.22}$$

$$\dot{\alpha} = \ddot{x}_d - k_1 \dot{z}_1 \tag{8.23}$$

The error dynamics of the second variable is defined as follows:

$$\dot{z}_2 = \dot{x}_2 - \dot{\alpha} = f + gu - \ddot{x}_d + k_1 z_2 - k_1^2 z_1 \tag{8.24}$$

The control law is selected as follows:

$$u = \frac{1}{g}\left(-f + \ddot{x}_d - k_1 z_2 + k_1^2 z_1 - k_2 z_2\right) \tag{8.25}$$

The values of functions g and f in the back-stepping control law in Eq. (8.25) are as follows:

$$g = 1$$
$$f = -0.4x_2 + 1.1x_1 - x_1^3 + 2.1\cos(1.8t)$$

The control law is rewritten as follows:

$$u = \left(0.4x_2 - 1.1x_1 + x_1^3 - 2.1\cos(1.8t) + \ddot{x}_d - k_1 z_2 + k_1^2 z_1 - k_2 z_2\right)$$

8.3.1 Stability Analysis

In this section, the stability of controller, designed in Section 8.3, is discussed based on the Lyapunov stability theorem. The proof of asymptotic stability is illustrated in the following theorem.

Theorem 8.1

For the chaotic Duffing system with (8.17) equations, the proof of global asymptotic stability is defined as follows. First, the stabilizing control law $x_2 = \phi(x_1)$ is found. This law is defined as follows:

$$\phi(x_1) = \alpha = -k_1 x_1, k_1 > 0 \tag{8.26}$$

The state variable x_2 is deleted.

$$\dot{x}_1 = -k_1 x_1 \tag{8.27}$$

Equation (8.27) is globally asymptotically stable. To guarantee the stability of Eq. (8.27), the candidate Lyapunov function is considered as follows. The proposed Lyapunov function must be positive definite.

$$v_1(x_1) = \frac{1}{2}x_1^2 \tag{8.28}$$

The derivative of the Lyapunov function must be negative definite.

$$\dot{v}_1 = x_1\dot{x}_1 = x_1[-k_1x_1] = -k_1x_1^2 \leq 0 \tag{8.29}$$

In the next part, variables are considered as follows:

$$z = x_2 - \phi(x_1) = x_2 + k_1x_1 \rightarrow \dot{x}_1 = x_2 = z - k_1x_1 \tag{8.30}$$

$$\dot{\phi}(x_1) = -k_1\dot{x}_1 \tag{8.31}$$

$$\dot{z} = \dot{x}_2 - \dot{\phi}(x_1) = f + gu + k_1(z - k_1x_1) \tag{8.32}$$

Now the new system is rewritten in terms of x_1 and z.

$$\begin{aligned} \dot{x}_1 &= z - k_1x_1 \\ \dot{z} &= f + gu + k_1(z - k_1x_1) \end{aligned} \tag{8.33}$$

Here, for system (8.33), a two-step Lyapunov function should be proposed, so that the global asymptotic stability of the system is guaranteed. The proposed Lyapunov function is semi-positive definite.

$$v(x_1, z) = v_1(x_1) + \frac{1}{2}z^2 = \frac{1}{2}x_1^2 + \frac{1}{2}z^2 = \frac{1}{2}\left(x_1^2 + z^2\right) \tag{8.34}$$

The derivative of the Lyapunov function is defined as:

$$\dot{v} = x_1\dot{x}_1 + z\dot{z} \tag{8.35}$$

$$\dot{v} = x_1[z - k_1x_1] + z[f + gu + k_1(z - k_1x_1)] = x_1z - k_1x_1^2 + z\left[f + gu + k_1z - k_1^2x_1\right]$$

$$\dot{v} = -k_1x_1^2 + z\left[x_1 + f + gu + k_1z - k_1^2x_1\right] \tag{8.36}$$

Now the back-stepping controller u is proposed to guarantee $\dot{v}$ is negative definite.

$$u = \left(-x_1 + 0.4x_2 - 1.1x_1 + x_1^3 - 2.1\cos(1.8t) - k_1z + k_1^2x_1 - k_2z\right); k_1, k_2 > 0 \tag{8.37}$$

Here, Eq. (8.37) is combined with Eq. (8.36). The resulting Eq. (8.37) is negative definite, and the global asymptotic stability of system (8.36) is guaranteed.

$$\dot{v} = -k_1x_1^2 - k_2z^2 \tag{8.38}$$

Equation (8.38) is negative definite for positive values of k_1, k_2 and it is nonzero except for the origin. Therefore, the designed controller makes the global asymptotic stability of the chaotic Duffing system.

Theorem 8.2
Similarly to prove asymptotic stability, the following Lyapunov function can be considered.

$$v_1 = \frac{1}{2} z_1^2 \tag{8.39}$$

The derivative of the Lyapunov function is as follows:

$$\dot{v}_1 = z_1 \dot{z}_2 = z_1 (z_2 - k_1 z_1) \tag{8.40}$$

$$\dot{v}_1 = z_1 z_2 - k_1 z_1^2 \Rightarrow z_1 (x_2 - \alpha) - k_1 z_1^2 \tag{8.41}$$

For $\dot{v}_1 \leq 0$, $\alpha = x_2$ must be chosen.
As a result

$$\dot{v}_1 = -k_1 z_1^2 \leq 0 \tag{8.42}$$

The second Lyapunov function is determined as follows:

$$v_2 = v_1 + \frac{1}{2} z_2^2 \tag{8.43}$$

The derivative of the second Lyapunov function is as follows:

$$\dot{v}_2 = \dot{v}_1 + z_2 \dot{z}_2 \tag{8.44}$$

$$\dot{v}_2 = -k_1 z_1^2 + z_2 \left(f + gu - \ddot{x}_d - k_1^2 z_1 \right)$$

For $\dot{v}_2 \leq 0$, u must be selected as follows:

$$u = \frac{1}{g} \left(-f + \ddot{x}_d - k_1 z_2 + k_1^2 z_1 - k_2 z_2 \right); k_1, k_2 > 0 \tag{8.45}$$

$$\dot{v}_2 = -k_1 z_1^2 - k_2 z_2^2 \leq 0 \tag{8.46}$$

By inserting Eq. (8.45) into Eq. (8.44), $\dot{v}_2$ becomes Eq. (8.46) and the Lyapunov function becomes negative definite except for the origin, so according to LaSalle Theorem system is asymptotically stable. As a result, the designed controller makes the chaotic Duffing oscillator system stable.

8.4 MATLAB Programming of Back-Stepping Controller

In this part, the simulation is done to check the performance of the designed controller in terms of tracking modes, chattering reduction, and reaching time. The MATLAB program for simulating the back-stepping control of Duffing system is as follows. The following program describes the equations of the chaotic Duffing system and the designed controller.

8.4.1 Back-Stepping Controller Design in MATLAB

In this section, the controller which was designed in Section 8.3 is applied on the chaotic Duffing oscillator system. Accordingly, we suppose $X_0 = [0.2,\ 0]$ and $X_d = \sin(t)$. X_0 is the initial condition and X_d is the input reference signal. The values of $k_1 = 2$ and $k_2 = 1.8$ are selected.

8.4.2 Controller Design Based on Back-Stepping Super-Twisting SMC

In recent years, many studies have been done to design the back-stepping controller based on the sliding mode and its developed methods [10]. One of these combined methods is combining back-stepping with super-twisting SMC. The super-twisting SMC method is fully discussed in Chapter 7. In this section, the controller is designed with back-stepping and super-twisting SMC methods for the nonlinear inverted pendulum system, with described equations of Section 7.2.1.

```
function [xdot]=Back-Steppingduf(t,x)
%defining the system parameters
p1 = 0.4; p2 = -1.1; p3 = 1; p4 = 2.1; w = 1.8;
k1 = 2; k2 = 1.8;

xr = sin(t);        %desired input signal
xrdot = cos(t); xrddot = -sin(t);

% Back-Stepping controller design
z1 = x(1) - xr;
alpha = -k1 * z1 + xrdot;
z2 = x(2)- alpha;
x1dot = x(2);
z1dot = x1dot - xrdot;
alphadot = -k1 * z1dot + xrddot;
u = 0.4*x(2)+(-1.1*x(1))+(x(1)^3)-2.1*cos(w*t)+(xrddot-
k1*z2+k1^2*z1-          k2*z2-z1);
dlmwrite('duffingcontrol.txt',u,'-append');
x2dot = (-p1*x(2)-p2*x(1)-p3*x(1)^3+p4*cos(w*t)+u);
% describing the statespace
```

```
xdot=[x1dot,x2dot]';
end
```

The next program calls and solves the dynamic equations of the chaotic Duffing system and finally draws the graphs related to the tracking of the reference signal by the state variables, the phase portrait, the tracking error, and the control signal.

```
clc
clear all
close all
tspan = [0 20]; init=[0.2,0];
delete('duffingcontrol.txt');
[T,y] = ode45('Back-Steppingduf',tspan,init);
ref = sin(T); refdot=cos(T);
error1 = abs(ref-y(:,1)); error2=abs(refdot-y(:,2));
%plotting of response without control effort
figure(1)
plot(T,sin(T),T,y(:,1),'--r','linewidth',3);
grid on
xlabel('t((sec)'); ylabel('amplitude'); title
('back-stepping controller ')
legend('reference signal','x1 state variable of the
system');
figure(2)
plot(T,cos(T),T,y(:,2),'--r','linewidth',3);
grid on
xlabel('t(sec)'); ylabel('amplitude'); title
('Back-Stepping controller ')
legend('reference signal','x2 state variable of the
system');
figure(3)
plot(y(:,1),y(:,2),'linewidth',3);
grid on
xlabel('Mode x1'); ylabel('Mode x2');
title('phase portrait');
figure(4)
plot(T,error1,'--k','linewidth',3);
grid on
xlabel('time(sec)');ylabel('amplitude');
title('Tracking error for state variable x1');
figure(5)
```

```
plot(T,error2,'--k','linewidth',3);
grid on
xlabel('time(sec)'); ylabel('amplitude');
title('Tracking error for state variable x2');
% plotting of control signal
u = dlmread('duffingcontrol.txt');
figure(6)
plot(T,u(1:237),'linewidth',3);
grid on
xlim([0 20])
ylim([-4 4])
xlabel('time (sec)'); ylabel('amplitude'); title('control
signal');
```

These diagrams are drawn in Figures 8.3–8.8.

With the controller designed by the back-stepping method, the state variables quickly reach the input reference signal and the error converges to zero. Also, the input control signal is in good condition and chattering phenomenon does not appear in it. The simulation of the chaotic Duffing system in the Simulink environment of MATLAB software is as shown in Figure 8.9.

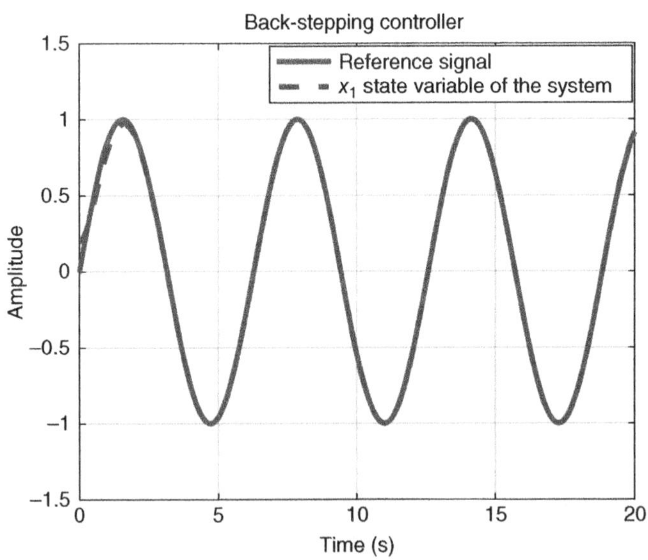

Figure 8.3 Tracking the reference input signal by the state variable x_1.

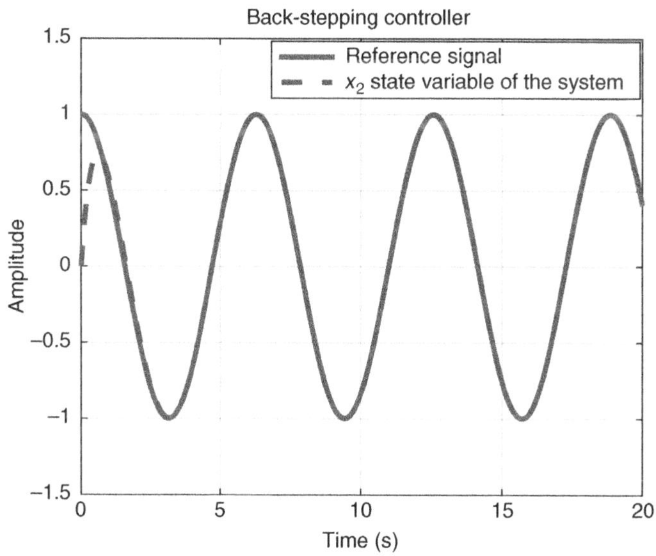

Figure 8.4 Tracking the reference input signal by the state variable x_2.

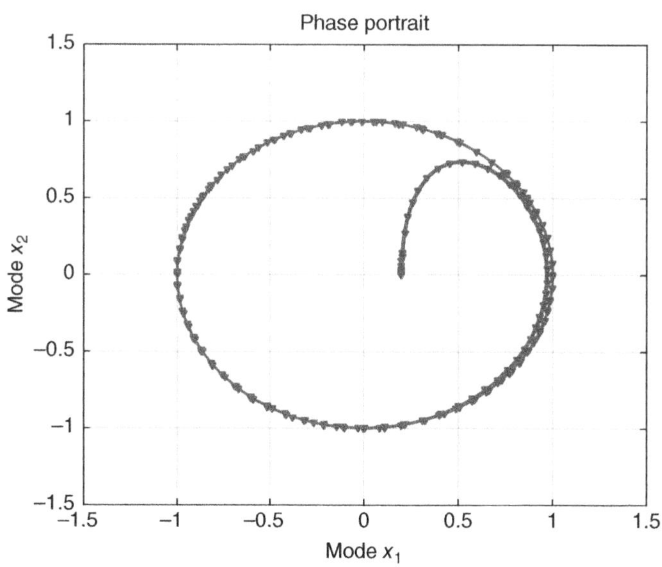

Figure 8.5 Phase portrait diagram.

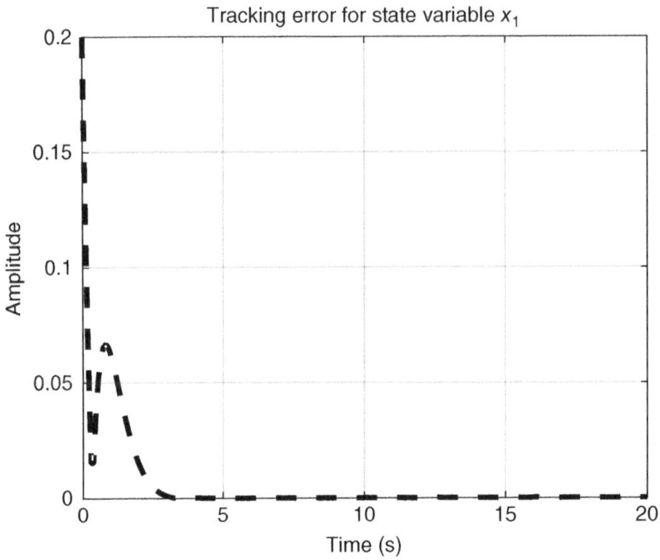

Figure 8.6 Tracking error related to x_1.

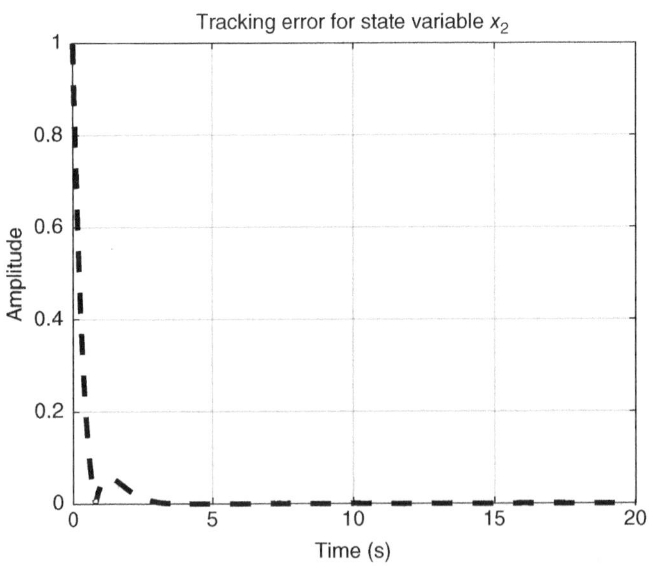

Figure 8.7 Tracking error related to x_2.

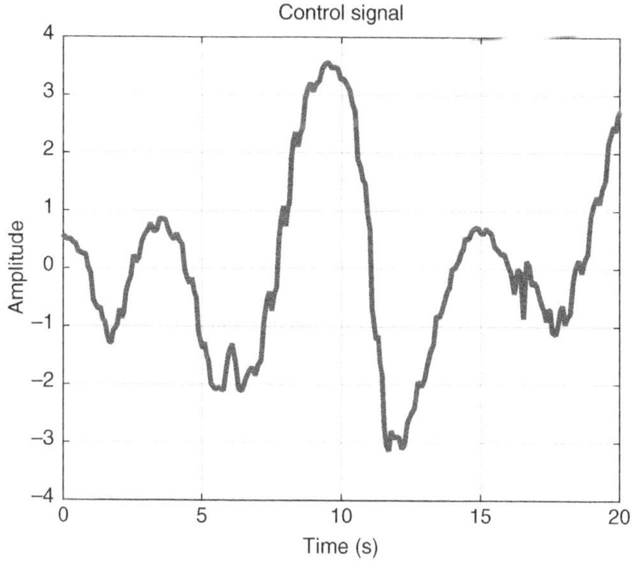

Figure 8.8 Input control signal.

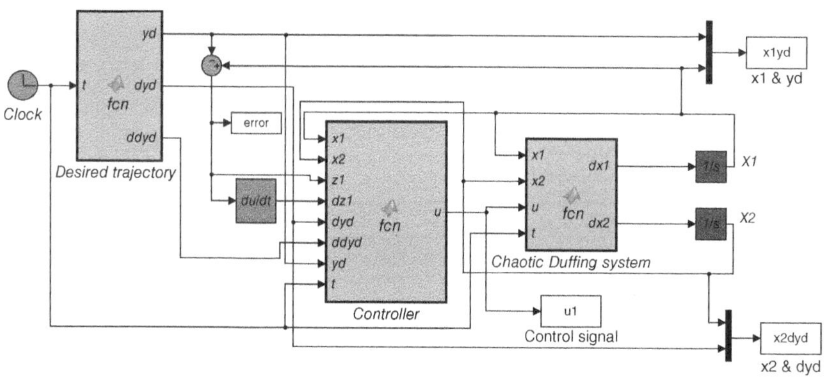

Figure 8.9 Block diagram of the system in Simulink.

The desired trajectory function program is as follows:

```
function [yd,dyd,ddyd] = fcn(t)
yd = sin(t);          % The desired input signal
dyd = cos(t);
ddyd = -sin(t);
```

The controller function program is as follows:

```
function u  = fcn(x1,x2,z1,dz1,dyd,ddyd,yd,t)
p1 = 0.4; p2 = -1.1; p3 = 1; p4 = 2.1; w = 1.8; k1 = 2;
k2 = 1.8;
alpha = -k1*z1+dyd;
z2 = x2 - alpha;
x1dot = x2;
dz1 = x1dot - dyd;
% control signal
u = p1*x2+(p2*x1)+(p3*x1^3)-p4*cos(w*t)+(ddyd-k1*z2
+k1^2*z1-k2*z2);
```

The function program of the Duffing chaotic system is as follows:

```
function [dx1,dx2]= fcn(x1,x2,u,t)
% System parameters
p1 = 0.4; p2 = -1.1; p3 = 1; p4 = 2.1; w = 1.8;
% Chaotic Duffing system equations
dx1 = x2;
dx2 = -p1*x2-p2*x1-p3*x1^3+p4*cos(w*t)+u;
```

The simulation results of the Simulink section are as in Figures 8.10–8.13.

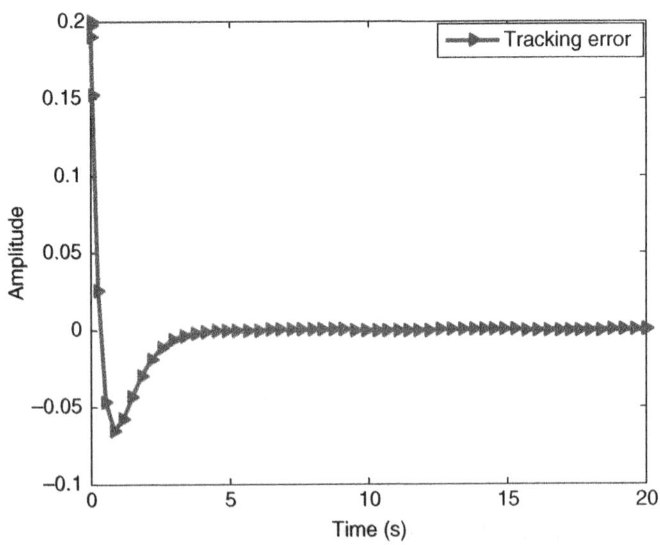

Figure 8.10 Tracking error implementing back-stepping controller in Simulink.

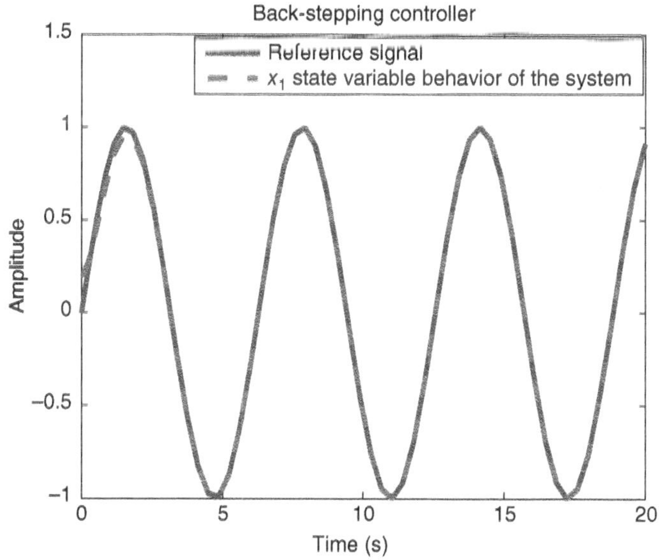

Figure 8.11 Reference tracking by x_1.

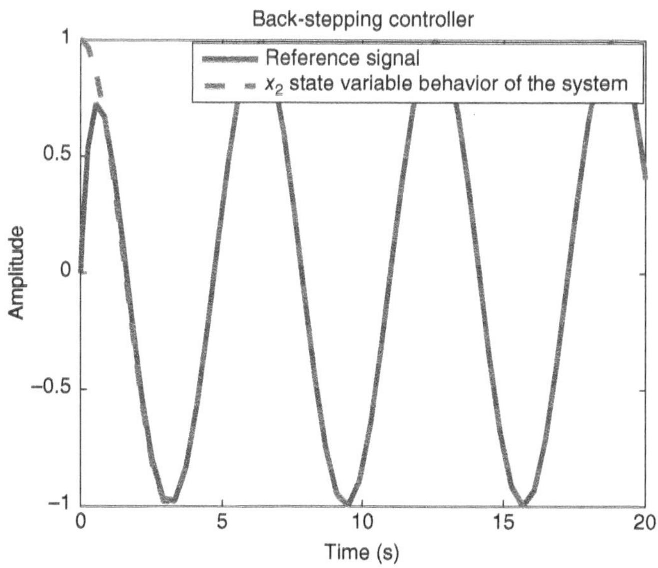

Figure 8.12 Reference tracking by x_2.

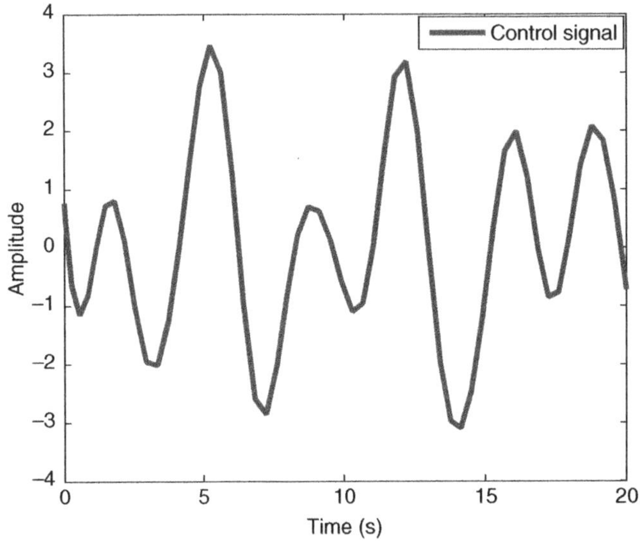

Figure 8.13 Input control signal in Simulink.

8.5 Back-Stepping Super-Twisting Sliding Mode Controller Design

Back-stepping control technique is widely used to achieve the goals of industrial nonlinear system control including military equipment, robotics, medical engineering, and technical services [11, 12]. The control law, equations, and parameters related to the super-twisting method are fully described in Chapter 7.

Theorem 8.3
Consider the dynamical inverted pendulum in Chapter 7 (Eq. 7.2). The stable back-stepping super-twisting SMC is designed as:

$$u = \frac{1}{g}(-f + c_1\dot{e}_1 + \ddot{x}_d + c_2 s + e_1 + k\ \mathrm{sign}(s)) \tag{8.47}$$

$$u = \frac{\frac{4}{3}(M+m)L - mL\cos^2 x_1}{\cos x_1}$$

$$\left(- \frac{(M+m)g\sin x_1 - mLx_2^2(\sin x_1 \cos x_1)}{\frac{4}{3}(M+m)L - mL\cos^2 x_1} + c_1\dot{e}_1 + \ddot{x}_d + c_2 s + k\ \mathrm{sign}(s) \right)$$

8.5.1 Proof of Stability

Consider the tracking error of state x_1 as follows.

$$e_1 = x_d - x_1, \quad \dot{e}_1 = \dot{x}_d - \dot{x}_1 = \dot{x}_d - x_2 \tag{8.48}$$

The positive definite Lyapunov function is defined as:

$$v_1 = \frac{1}{2}e_1^2 \tag{8.49}$$

And its related derivative is:

$$\dot{v}_1 = e_1\dot{e}_1 = e_1(\dot{x}_d - x_2) \tag{8.50}$$

To guarantee $\dot{V}_1 \leq 0$, x_2 is defined as follows:

$$x_2 = -s + c_1e_1 + \dot{x}_d \tag{8.51}$$

Considering the above relation, sliding surface is proposed as follows:

$$s = c_1e_1 + \dot{x}_d - x_2 = c_1e_1 + \dot{e}_1 \quad c_1\rangle 0 \tag{8.52}$$

Replacing (8.51) in (8.50) yields:

$$\dot{v}_1 = e_1(\dot{x}_d - (-s + c_1e_1 + \dot{x}_d)) = e_1s - c_1e_1^2 \tag{8.53}$$

Consequently (8.54) results by deriving from (8.52)

$$\dot{s} = c_1\dot{e}_1 + \ddot{e}_1 = c_1\dot{e}_1 + \ddot{e}_1 = c_1\dot{e}_1 + \ddot{x}_d - \dot{x}_2 \tag{8.54}$$

Where (8.54) is rewritten in the form of (8.55):

$$\dot{s} = c_1\dot{e}_1 + \ddot{x}_d - f - gu \tag{8.55}$$

$$\dot{s} = c_1\dot{e}_1 + \ddot{x}_d - \frac{(M+m)g\sin x_1 - mLx_2^2(\sin x_1 \cos x_1)}{\frac{4}{3}(M+m)L - mL\cos^2 x_1} - \frac{\cos x_1}{\frac{4}{3}(M+m)L - mL\cos^2 x_1}u$$

In the second step, Lyapunov function v_2 is defined as follows:

$$v_2 = v_1 + \frac{1}{2}s^2 \tag{8.56}$$

Deriving (8.56), u appeared and should be denoted such that it satisfies $\dot{v}_2 \leq 0$

$$\dot{v}_2 = \dot{v}_1 + s\dot{s} = e_1s - c_1e_1^2 + s(c_1\dot{e}_1 + \ddot{x}_d - f - gu) \tag{8.57}$$

Accordingly u will be calculated by relation (8.58):

$$u = \frac{1}{g}(-f + c_1\dot{e}_1 + \ddot{x}_d + c_2 s + e_1 + k\ \mathrm{sign}(s)) \tag{8.58}$$

$$u = \frac{\frac{4}{3}(M+m)L - mL\cos^2 x_1}{\cos x_1}$$

$$\left(-\frac{(M+m)g\sin x_1 - mLx_2^2(\sin x_1 \cos x_1)}{\frac{4}{3}(M+m)L - mL\cos^2 x_1} + c_1\dot{e}_1 + \ddot{x}_d + c_2 s + k\ \mathrm{sign}(s)\right)$$

$$\Rightarrow \dot{v}_2 = -c_1 e_1^2 - c_2 s^2 - k|s| \le 0$$

The values $c_1 = c_2 = 5$ are selected.

8.5.2 MATLAB Programming

In this section, simulation is done to check the performance of the designed controller in terms of tracking errors, chattering, and reaching time. The simulation of the above system by the Simulink section in MATLAB software is shown in Figures 8.14 and 8.15. Since three control methods such as back-stepping super-twisting SMC, feedback linearization, and super-twisting SMC are compared in this section, the Simulink block functions for plotting the simultaneous graphs are shown in Figure 8.15. This program includes three subsystems.

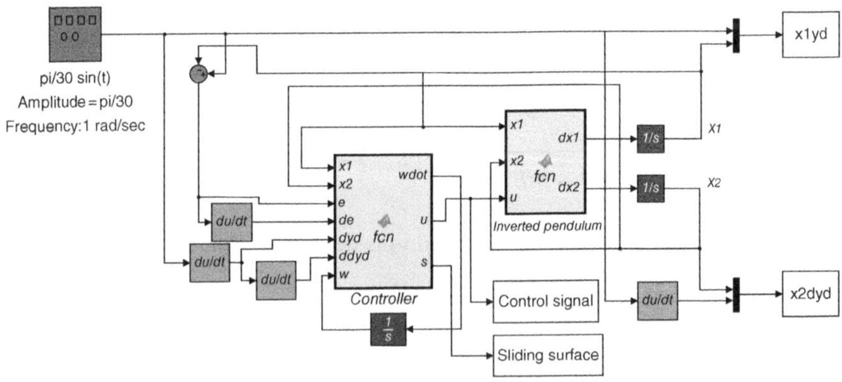

Figure 8.14 Block diagram of the system in Simulink.

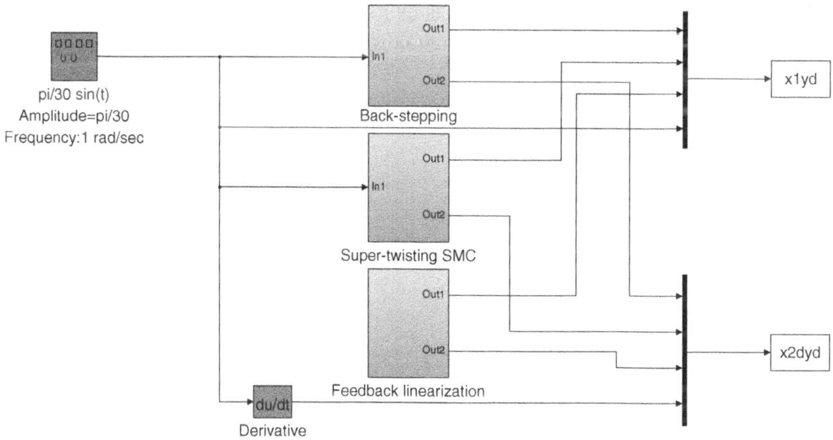

Figure 8.15 Block diagram of back-stepping super-twisting in Simulink.

8.5.3 Back-Stepping Super-Twisting SMC Design in MATLAB

In this section, the controller designed in Section 8.5 is applied on the inverted pendulum system. Accordingly we suppose $X_0 = [0.2, 0.1]$ and $X_d = \frac{\pi}{30}\sin(t)$. X_0 is the initial condition and X_d is the input reference signal.

The inverted pendulum function program is as follows:

```
function [dx1,dx2]= fcn(x1,x2,u)
m = 0.1;
M = 1;
L = 0.5;
g = 9.8;
zeros(2,1);
dx1 = x2;
f = ((M+m)*g*sin(x1)-m*L*(x2^2)*sin(x1)*cos(x1))/((4/3)*
(M+m)*L-m*L*(0.5+0.5*cos(2*x1)));
b = cos(x1)/((4/3)*(M+m)*L-m*L*(0.5+0.5*cos(2*x1)));
dx2 = f+b*u;
```

The input signal function program is as follows:

```
function [wdot,u,s] = fcn(x1,x2,e,de,dyd,ddyd,w)

m = 0.1;
M = 1;
L = 0.5;
g = 9.8;
k = 18;
EPS = 0.01;
y = 5;
c1 = 5;
c2 = 5;
s = de+y*e%u1 = -k*sign(s);
%u1 = -k*s/(abs(s)+EPS)%u1 = -k*tanh(s/0.5)%u1 = -k*SAT(s);
u1 = -2*(abs(s))^0.5*sign(s)+w % Super twisting
controlwdot = -8*sign(s);
f = ((M+m)*g*sin(x1)-m*L*(x2^2)*sin(x1)*cos(x1))/((4/3)*
(M+m)*L-m*L*(0.5+0.5*cos(2*x1)));
b = cos(x1)/((4/3)*(M+m)*L-m*L*(0.5+0.5*cos(2*x1)));
u = (1/b)*(-f+ddyd+c1*de+c2*s+e-u1);
%end
%function Z = SAT(S) lta = 0.1
%if S > delta
%Z = 1;
%elseif S<-delta
%Z = -1;
%else
%Z = 1/delta*S;
%end
%end
```

Comparing Figure 8.16 with Table 8.1, it is clear that back-stepping super-twisting SMC contains less surge in specified points.

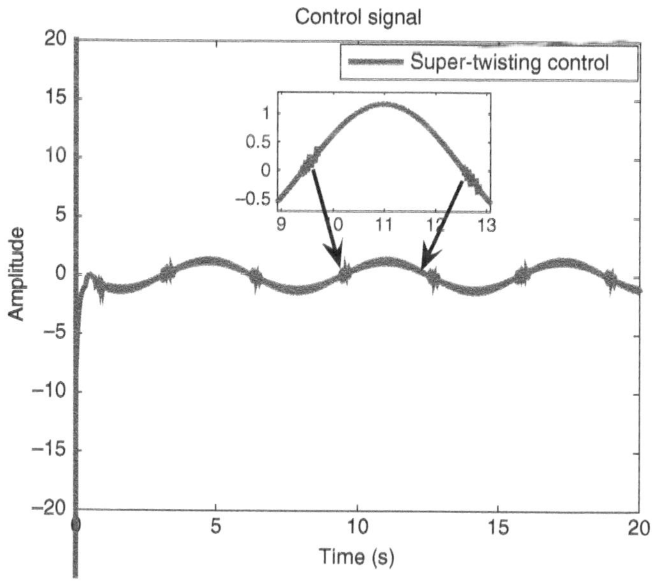

Figure 8.16 Control signal in back-stepping super-twisting SMC.

Table 8.1 Performance analysis of control strategies.

Response specification	Steady-state error	Convergence time (s)	Over shoot (u)	The peak-to-peak amplitude of the u signal	MSE
Feedback linearization	0.0003	<8	1.23	2.48	0.0064
Super-twisting SMC	0	<2	1.17	2.35	0.0046
Back-stepping super-twisting	0	<1.4	1.17	2.35	0.0036

Additionally, comparative results of Table 8.2 and Figure 8.17 show better convergence of sliding surface to zero in the proposed method.

Considering Figures 8.18 and 8.19, it is clear that proposed method performs less convergence time. Finally, Figure 8.20 indicates the tracking errors.

Table 8.1 represents a brief comparison between feedback linearization, super-twisting SMC, and back-stepping super-twisting.

Table 8.2 Comparison of different methods in terms of the convergence time.

Response specification	Back-stepping super-twisting	Super-twisting SMC
Convergence time to sliding surface	1S	2S

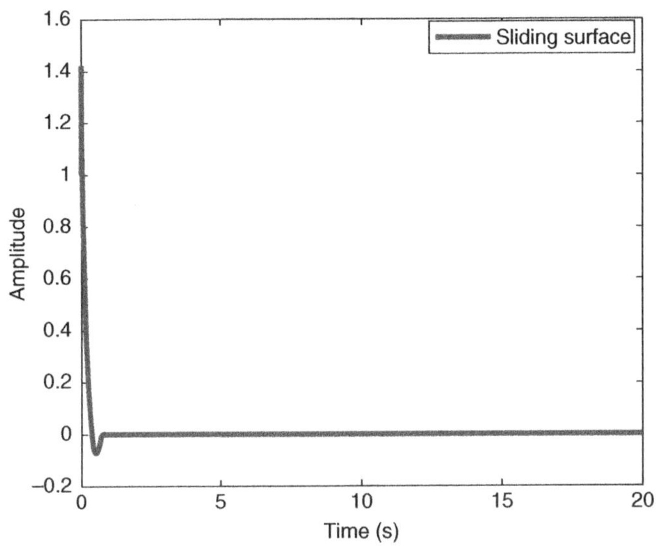

Figure 8.17 Sliding surface in back-stepping super-twisting method.

According to Table 8.1, in terms of the accuracy and states' convergence speed, the back-stepping super-twisting SMC method shows a very good performance and exactly converges to the reference signal. In addition, it not only reduces the overshoot value but is also able to control and eliminate the chattering phenomenon. According to MSE values, it is concluded that the back-stepping super-twisting method has the lowest MSE value compared to the other methods mentioned in Table 8.1.

According to Figure 8.21, the state variables in the back-stepping super-twisting SMC method reach the sliding surface in less than 1 s and remain on sliding surface. This indicates the efficiency of the control signal, as illustrated in Table 8.2 and Figure 8.21.

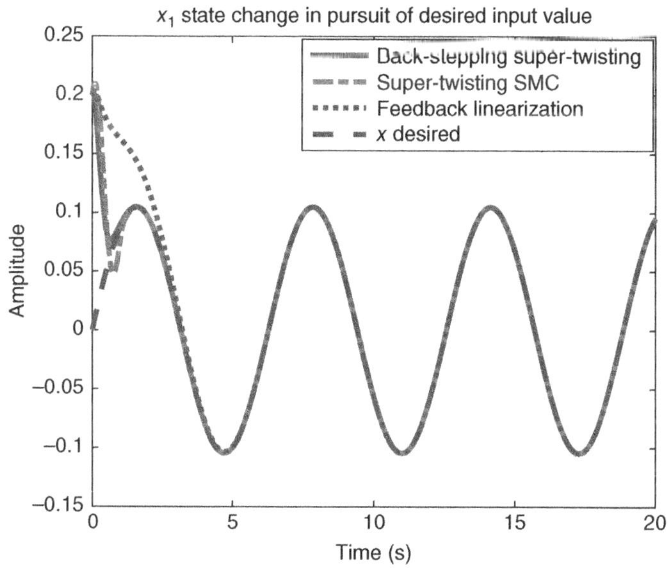

Figure 8.18 Comparative results of reference tracking by x_1.

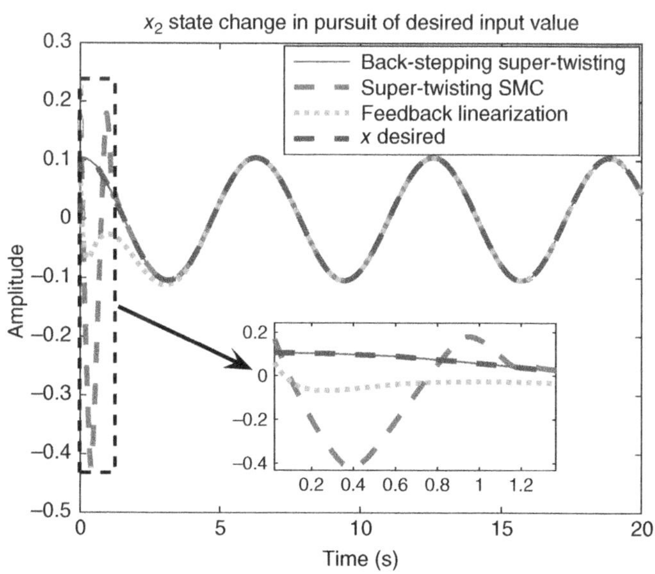

Figure 8.19 Comparative results of reference tracking by x_2.

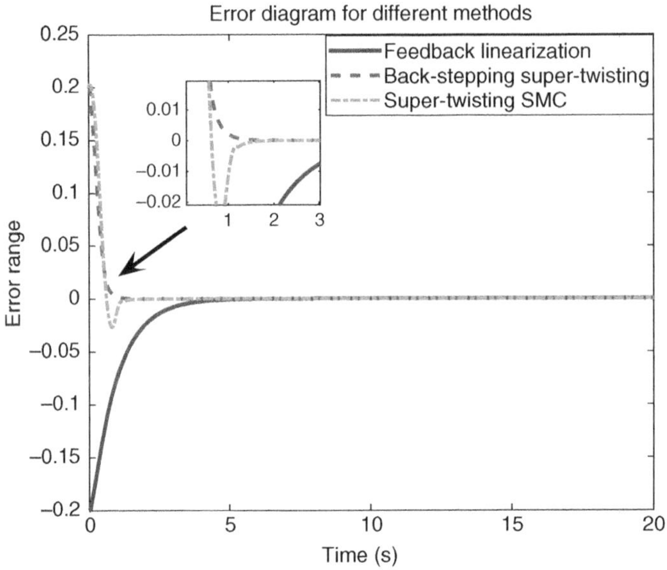

Figure 8.20 Comparative results of tracking errors.

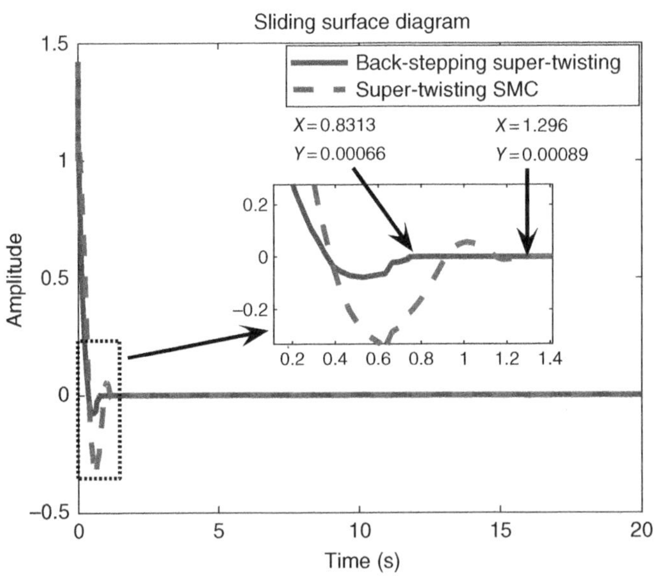

Figure 8.21 Sliding surface in back-stepping super-twisting SMC method and super-twisting SMC method.

In Table 8.2, super-twisting SMC and back-stepping super-twisting SMC methods are compared in terms of convergence time to sliding surface with an error less than 0.0001. As can be seen in Table 8.2, in the back-stepping super-twisting SMC method, the state variables have reached the sliding surface very quickly.

References

1 Kokotovic, P.V. (1992). The joy of feedback: nonlinear and adaptive. *IEEE Control Systems Magazine* 12 (3): 7–17.

2 Lozano, R. and Brogliato, B. (1992). Adaptive control of robot manipulators with flexible joints. *IEEE Transactions on Automatic Control* 37 (2): 174–181.

3 Khalil, H.K. (2002). *Nonlinear Systems*, 2e. Prentice Hall.

4 Ali, S., Prado, A., and Pervaiz, M. (2023). Hybrid Back-Stepping-Super twisting algorithm for robust speed control of a three-phase induction motor. *Electronics* 12: 681.

5 Hassan, S., Abdelmajid, B., Mourad, Z. et al. (2020). PSO-Backstepping controller of a grid connected DFIG based wind turbine. *International Journal of Electrical and Computer Engineering (IJECE)* 10 (1): 856–867.

6 Kabziński, J. (2010). *Adaptive Control of Duffing Oscillator with Unknown Input Gain*. IEEE.

7 Marquez, H.J. (2003). *Nonlinear Control Systems*. Wiley.

8 Ge, S.S., Wang, C., and Lee, T.H. (2000). Adaptive back-stepping control of a class of chaotic systems. *International Journal of Bifurcation and Chaos* 10 (5): 1149–1156.

9 Karimov, A., Rybin, V., Dautov, A. et al. (2023). Mechanical chaotic duffing system with magnetic springs. *Inventions* 8: 19.

10 Zhou, J. and Wen, C. (2008). *Adaptive Backstepping Control of Uncertain Systems*. Springer.

11 Gong, X., Wen, F., Bian, X., and Fei, J. (2023). Adaptive back-stepping terminal sliding mode control of nonlinear system using fuzzy neural structure. *Mathematics* 11: 1094.

12 Almakhles, D.J. (2020). Robust Backstepping sliding mode control for a quadrotor trajectory tracking application. *IEEE ACESS* 8. https://doi.org/10.1109/ACCESS.2019.2962722.

9

Chattering Reduction Methods

9.1 Introduction

Chattering phenomenon has been one of the most important controlling challenges in recent years, and efforts have been made to eliminate or control this phenomenon in the best possible way with various methods and control strategies. In practical applications, engineers have difficulty in implementing the sliding mode controller due to the chattering phenomenon, and this phenomenon damages industrial equipment, including valves and mechanical equipment, and due to this disadvantage, it is the most important obstacle for implementation in practical applications. When the state variables reach the sliding level, they start to oscillate and this behavior is not acceptable for a control signal applied to industrial equipment. To control and eliminate the chattering phenomenon, soft functions can be used instead of the sign function in the sliding mode control. In this chapter, different types of soft functions have been introduced and simulated to remove the chattering phenomenon. The results clearly show the effectiveness of the proposed control strategy not only in terms of stability and convergence improvement but also in reducing unwanted chatter. In this chapter, Section 9.1 introduces chattering phenomenon and its damages, and emphasizes the need to eliminate or reduce it. Section 9.2 introduces methods to eliminate or reduce chattering phenomenon. Finally, in Section 9.3, MATLAB programming and implementation of the sliding mode control method is discussed. The model discussed in this chapter is a nonlinear and highly unstable system of the inverted pendulum. Results of applying proposed controller on inverted pendulum are compared in terms of the tracking speed, convergence time, error, and chattering reduction.

Methods of Developing Sliding Mode Controllers: Design and MATLAB Simulation, First Edition.
Reihaneh Kardehi Moghaddam and Mostafa Rabbani.

Published 2025 by John Wiley & Sons, Inc.

9.2 Methodology

The simulation results in this chapter are considered by implementing all methodologies of chattering reduction on an inverted pendulum.

9.2.1 Methods to Reduce Chattering Phenomenon

One limitation of SMC in industrial implementation is the presence of unwanted and annoying chattering fluctuations which cause increasing control effort, inducing unmodeled high-frequency dynamics, reducing control accuracy, increasing heat losses in electrical circuits, and energy consumption in mechanical parts and actuators. The existence of the undesirable chattering phenomenon in the first-order sliding model control provides motivations for design methods to improve the behavior of the first-order sliding model control. As seen in Figure 9.1, when the state variables reach the sliding surface, they begin to oscillate, and these fluctuations lead to various problems in dynamic systems. In Table 9.1, some methodologies are introduced to remove or deal with chattering phenomenon. The types of functions examined in this chapter to address chattering issue are described in Figure 9.2. These methods are fully simulated and compared. In Section 9.2.2, the sliding surfaces are discussed using the methods mentioned in Table 9.1.

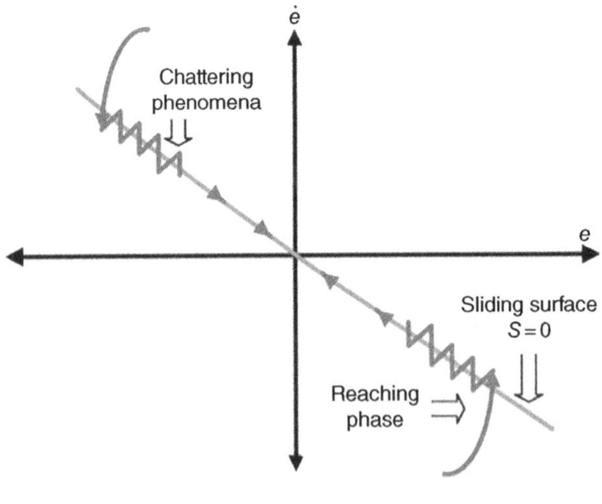

Figure 9.1 Fluctuation of state variables on the sliding surface.

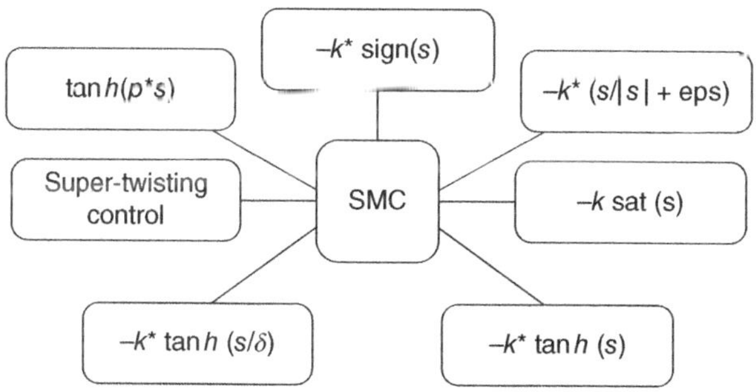

Figure 9.2 Proposed functions to eliminate the chattering phenomenon.

Table 9.1 Literature review to reduce chattering.

Reference	Variable structure part	Details
[1–9]	$U_{SW} = -K\,\text{sign}(s)$	$\text{sign}(s)\begin{cases} -1, s<0 \\ 0, s=0 \\ 1, s>0 \end{cases}$
[10]	$U_{SW} = -\text{Kep\_sign}(s)$	$\text{sign}(s) = \dfrac{s}{\lvert s\rvert + \varepsilon}$
[7, 8]	$U_{SW} = -\text{Ksat}(s)$	$\text{sat}(s) = \begin{cases} -1, s < -\phi \\ \dfrac{s}{\phi}, -\phi \le s \le \phi \\ 1, s>0 \end{cases}, 0<\phi<1$
[11]	$U_{SW} = -K\tan h(s)$	
[6, 9, 12]	$U_{SW} = -K\tan h\left(\dfrac{S}{\delta}\right)$	$\delta = 0.5$
[3, 6]	$U = -K\tan h(\rho s),\ K>0$	$\tan h(\rho s) = \dfrac{e^{\rho s} - e^{-\rho s}}{e^{\rho s} + e^{-\rho s}}, \rho > 0$
[8, 13]	$U_{ST} = -\alpha\lvert S\rvert^{P}\,\text{sign}(s) + U_1$ $\dot{U}_1 = -\omega\,\text{sign}(s)$	$\alpha = 2,\ p = 0.5,\ \omega = 8$

9.2.2 Sign Function

The results of the control signal and the sliding surface obtained from the function $U_{SW} = -K\,\text{sign}(s)$ are shown in Figure 9.3. As seen in Figure 9.3, the chattering phenomenon appears when the sign is used. Fluctuations on the sliding surface are also visible.

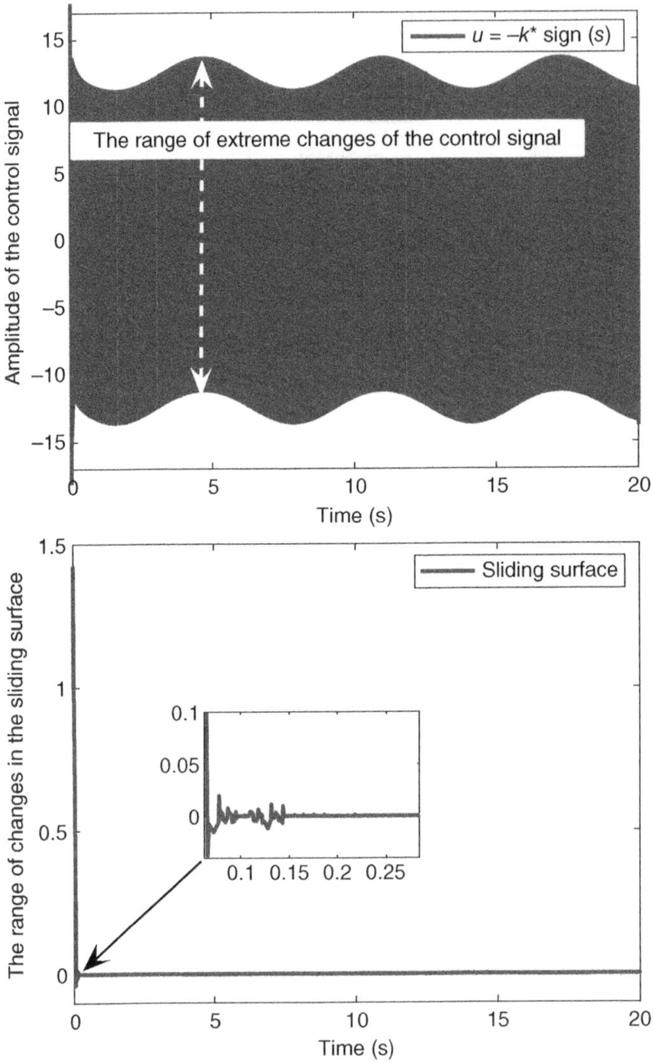

Figure 9.3 Control signal and sliding surface by function $U_{SW} = -K\,\mathrm{sign}(s)$.

9.2.3 Epsilon-Sign Function

The results of the control signal and the sliding surface obtained from the function $U_{SW} = -Kep\_\,\mathrm{sign}(s)$ are shown in Figure 9.4. Using the Ep-sign function, the chattering phenomenon was reduced, but not completely omitted. Also, when the system variables reach the sliding level, they start to fluctuate.

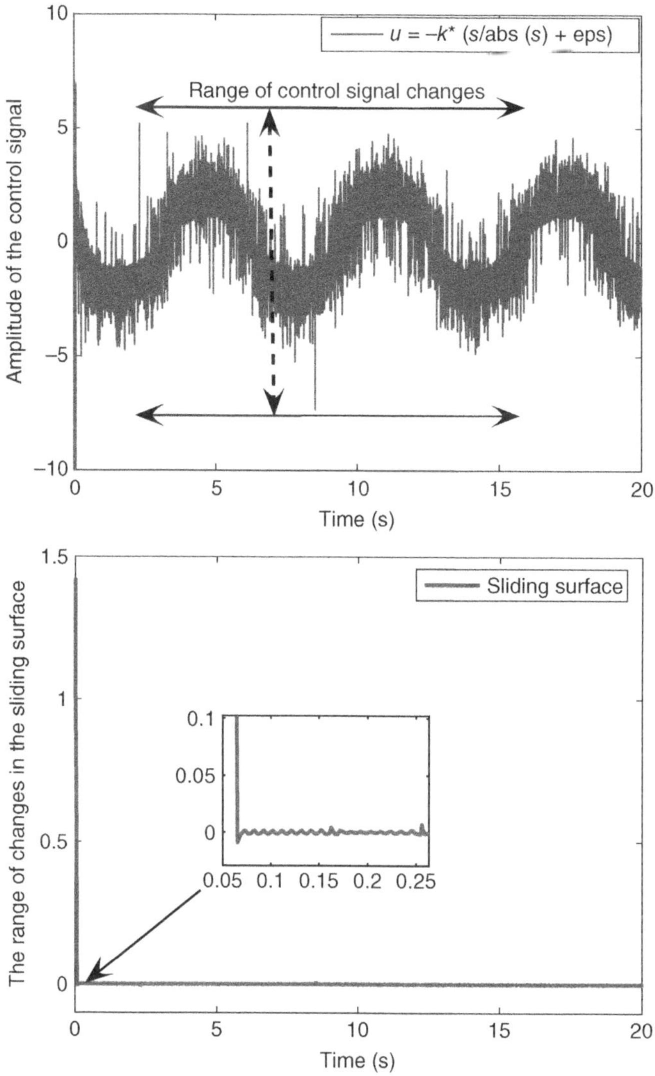

Figure 9.4 Control signal and sliding surface by function $U_{SW} = -K\text{ep\_sign}(s)$.

9.2.4 Saturation Function

The results of the control signal and the sliding surface obtained from the function $U_{SW} = -K\text{sat}(s)$ are shown in Figure 9.5. Although the use of the saturation function effectively eliminates the chattering phenomenon, there are still fluctuations in the initial moments. As it is illustrated, the oscillations were removed from the sliding surface.

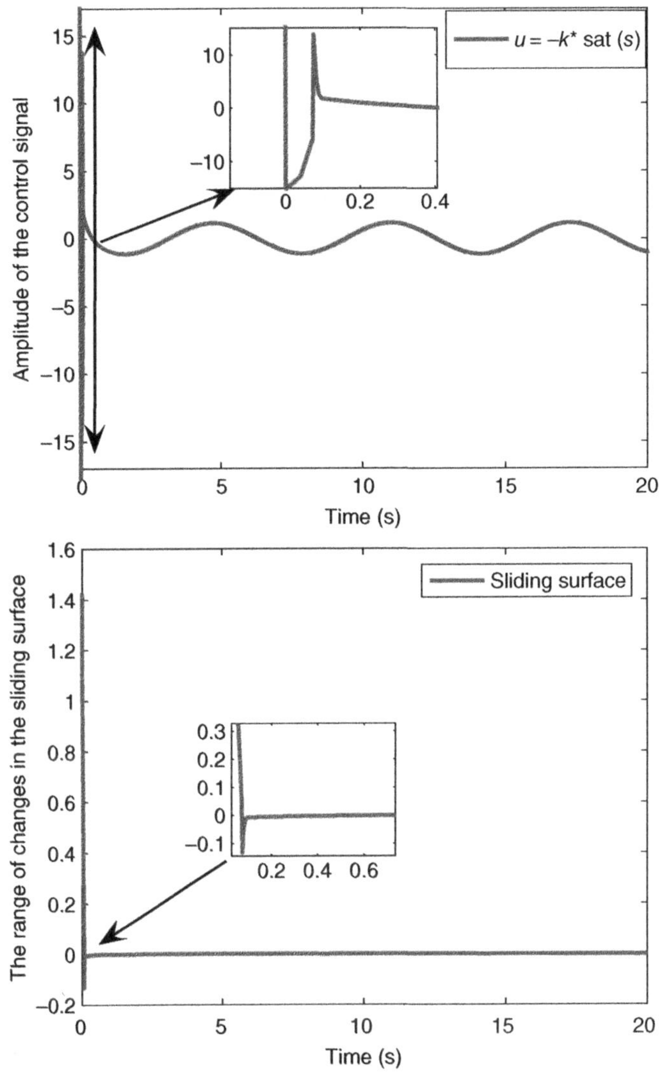

Figure 9.5 Control signal and sliding surface by function $U_{SW} = -K\text{sat}(s)$.

9.2.5 Hyperbolic Tangent Function

The results of the control signal and the sliding surface obtained from the function $U_{SW} = -K \tan h(s)$ are shown in Figure 9.6. Here, the control signal and the sliding surface have minor fluctuations.

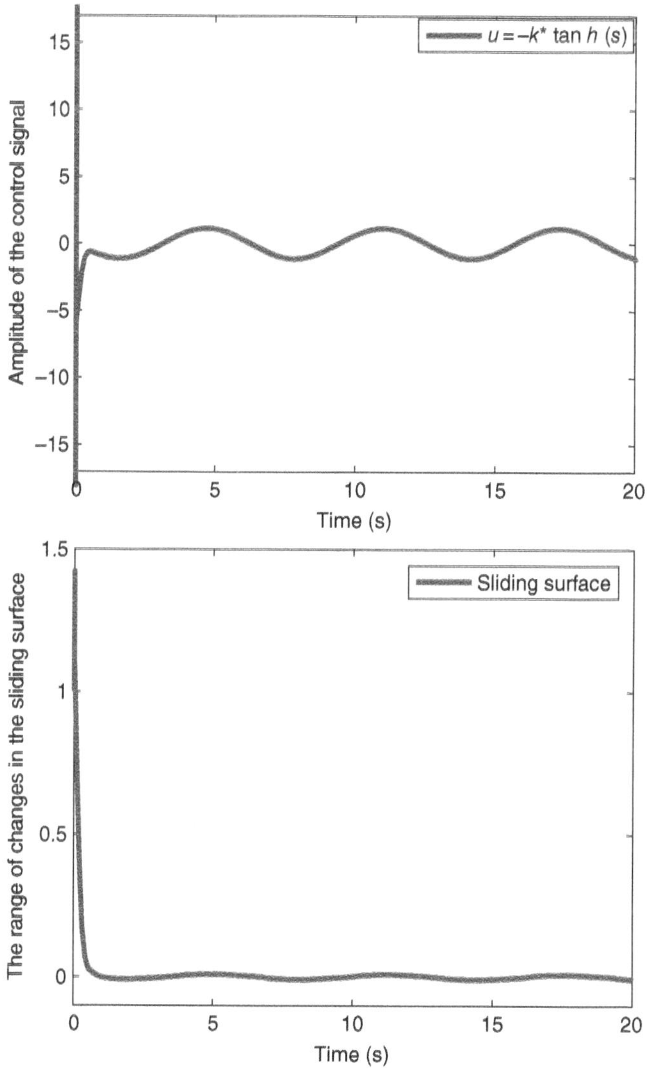

Figure 9.6 Control signal and sliding surface by function $U_{SW} = -K \tan h(s)$.

9.2.6 Generalized Hyperbolic Tangent Function: $U_{SW} = -K \tan h(S/\delta)$

The results of the control signal and the sliding surface obtained from the function $U_{SW} = -K\tan h\left(\frac{S}{\delta}\right)$ are shown in Figure 9.7. Using the generalized hyperbolic tangent function affected the sliding surface.

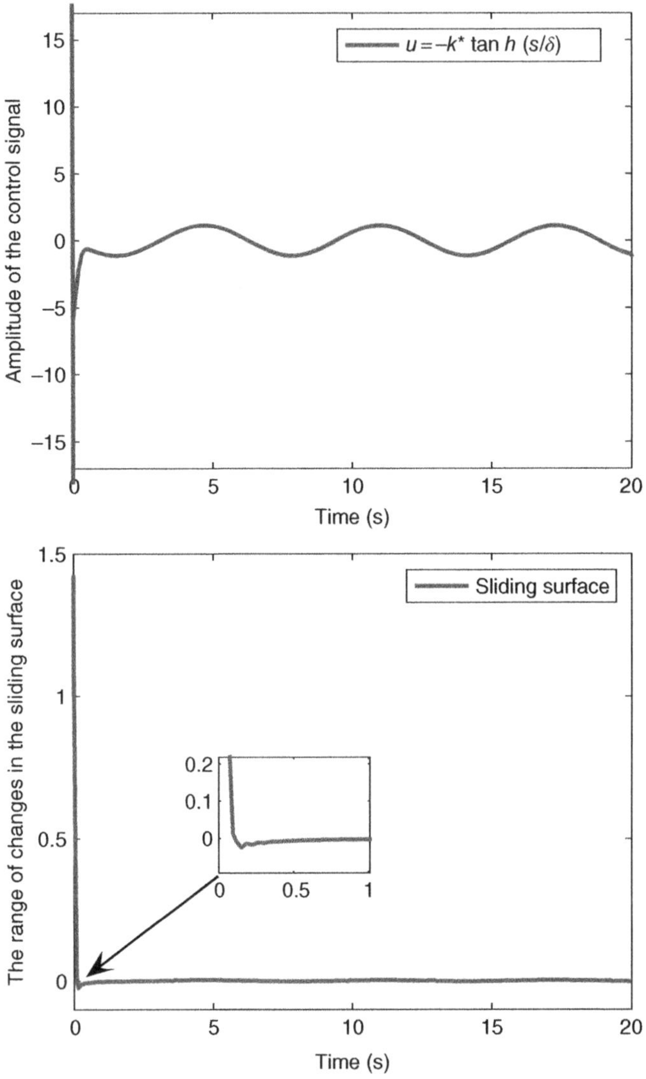

Figure 9.7 Control signal and sliding surface by function $U_{SW} = -K\tanh\left(\frac{S}{\delta}\right)$.

9.2.7 Generalized Hyperbolic Tangent Function: $-K \tan h(\rho s)$

The results of the control signal and the sliding surface which are related to the function $-K \tan h(\rho s)$ are shown in Figure 9.8. The state variables of the system are separated from the sliding surface when they have reached the sliding surface.

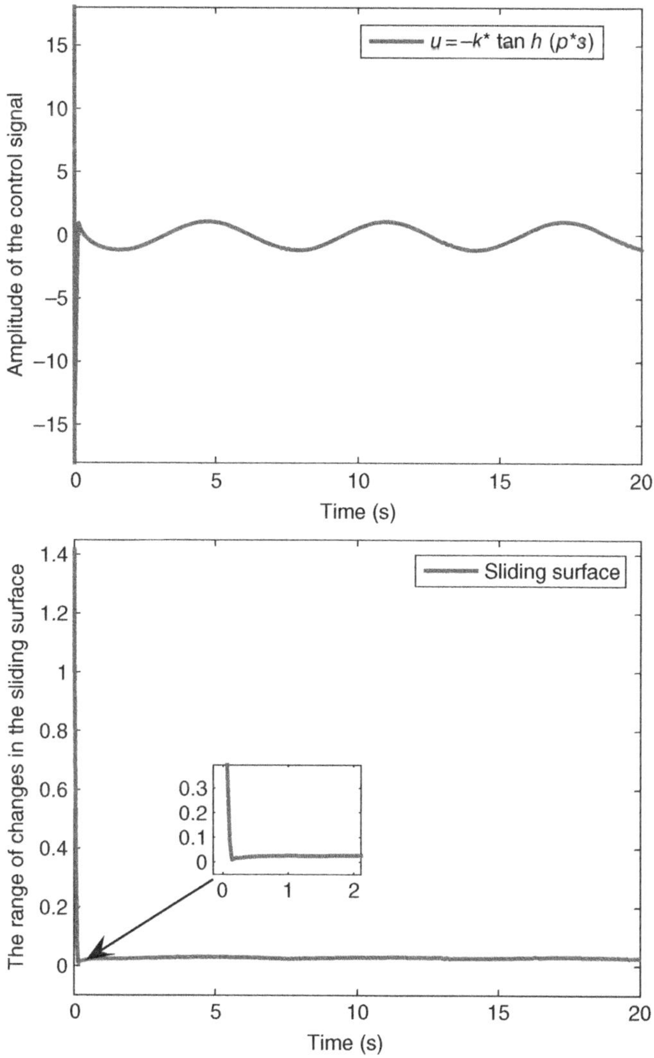

Figure 9.8 Control signal and sliding surface by function $U_{SW} - K \tan h(\rho s)$.

9.2.8 Super-Twisting

This method is fully explained in Chapter 7. Super-twisting algorithm, like the first-order sliding model control method, is a robust method for controlling systems in the presence of external disturbances with uncertainty. This algorithm is one of the most effective methods to improve the first-order sliding model control behavior and reduce chattering. The results of the control signal and the

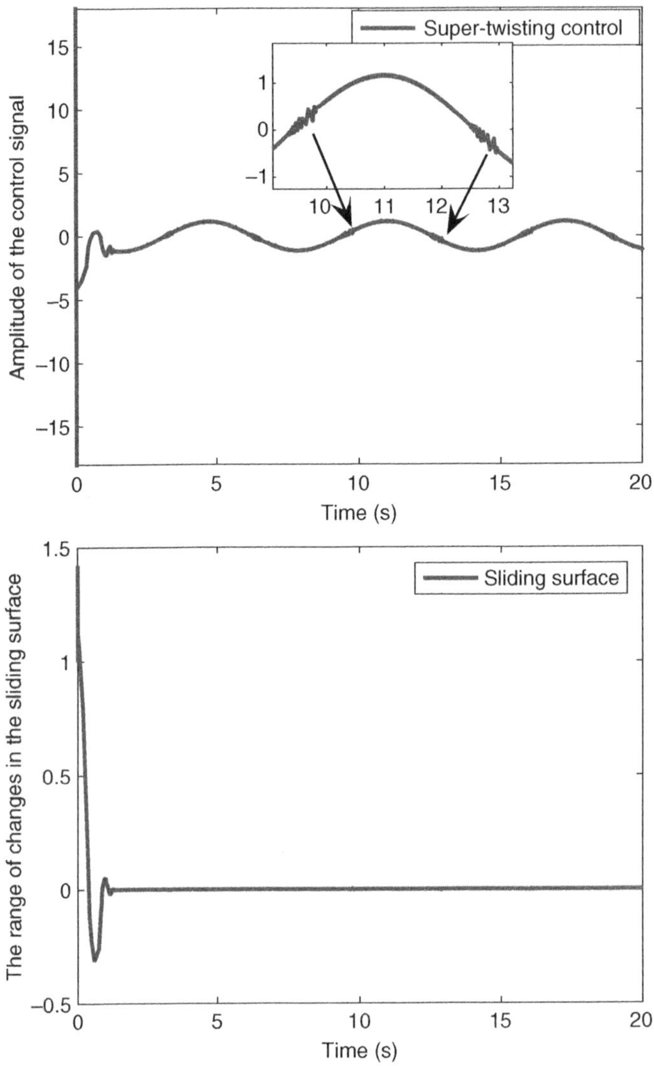

Figure 9.9 Control signal and sliding surface by function super-twisting.

sliding surface obtained from the function super-twisting are shown in Figure 9.9. Using this method, the system-state variables are attracted to the sliding surface at a suitable speed and are not separated.

As seen in above section, using the sign function causes drastic changes in the control signal, which causes damage to industrial equipment, and improved controlling signals greatly reduces chattering. In Table 9.2, the methods presented in

Table 9.2 Comparison of different methods in terms of the speed of reaching and absorbing the sliding surface.

Response specification	$-K$ sign(s)	$-K$ ep_ sign(s)	$-k$ sat(s)	$-K$tan $h(s)$	$-K\tanh\left(\frac{S}{\delta}\right)$	$-K$tan $h(\rho s)$	Super-twisting
Convergence time to sliding surface	5s	10s	10s	>20s	20s>	20s>	2s
Maximum error value after reaching time	0.0003	3.51E-5	3.96E-5	0.0073	0.00048	0.0261	3.038E

Table 9.1 are compared in terms of the time to converge to the sliding surface and the maximum tracking error after reaching the sliding surface.

According to Table 9.2, the state variables in super-twisting SMC method reach the sliding surface in less than 2 s and are attracted to this surface. In the following, it is shown that the absorption of the state variables to the sliding surface is effective in the operation of following the reference signal, and there is less error between the values of the state variables and the reference signal. Figures 9.10 and 9.11 show the performance of various functions and the sliding mode control method and how the system modes act in pursuit of the desired input signal.

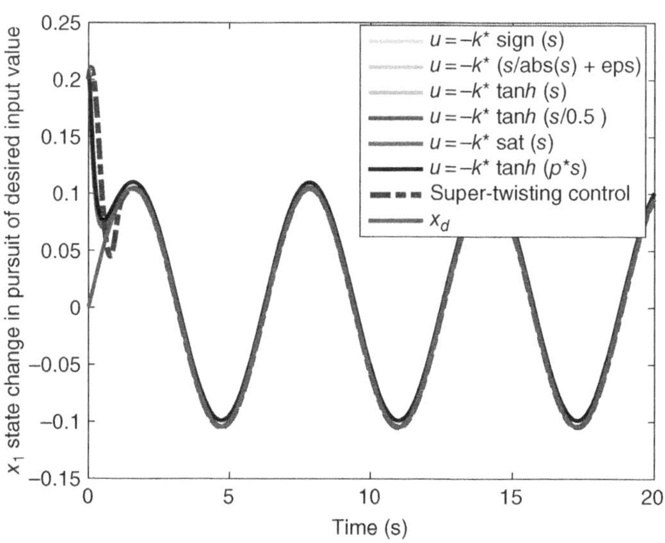

Figure 9.10 Reference tracking by x_1, using controlling structures of Table 9.1.

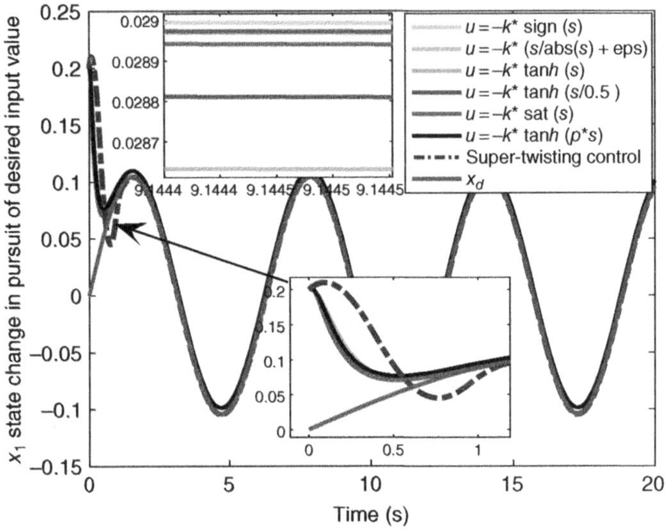

Figure 9.11 Reference tracking by x_1, using controlling structures of Table 9.1.

In Figures 9.10 and 9.11, the lowest amount of error and the best accuracy is related to the super-twisting control method, which completely and with negligible error follows the desired input signal by x_1. Of course, in terms of removing the chattering phenomenon, some other methods also showed good performance. To describe the performance of the above methods more precisely, in Table 9.3, a

Table 9.3 Performance analysis of control strategies.

Response specification	Steady-state error	Time to reach the desired value (s)	Over shoot (u)	The peak-to-peak amplitude of the u signal	MSE
$U_{SW} = -K\,\text{sign}(s)$	0.0001	<2	13.59	27.18	$6.690*10^{-4}$
$U_{SW} = -K\ \text{ep\_sign}(s)$	0.00008	<2	5.19	12.53	$6.7*10^{-4}$
$U_{SW} = -k\ \text{sat}(s)$	0.0001	<2	**1.163**	**2.326**	$6.68*10^{-4}$
$U_{SW} = -K\tan h(s)$	0.0006	<2.5	**1.14**	**2.28**	$6.81*10^{-4}$
$U_{SW} = -K\tan h\left(\frac{S}{\delta}\right)$	0.0004	<2.5	**1.146**	**2.29**	$6.690*10^{-4}$
Super_Twisting	**0**	<2	**1.176**	**2.351**	$9.9*10^{-4}$
$U = -K\tan h(\rho s)$	0.0052	20>	**1.091**	**2.269**	$7.3*10^{-4}$

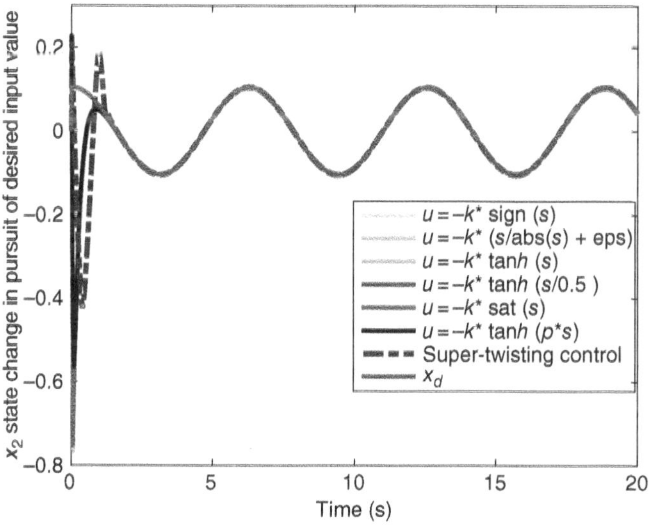

Figure 9.12 Reference tracking by x_2, using controlling structures of Table 9.1.

functional analysis of the control method is reviewed in terms of steady-state error, reaching time, the amount of overshoot of the control signal, the chattering range, and the MSE error, where the parameter N is the number of sampled points.

$$MSE = \frac{1}{N}\sum_{i=1}^{n}(X_i - X_{di})^2;\ X_i = X\,(i\,\Delta\,t) \tag{9.1}$$

According to Table 9.3, in terms of accuracy and convergence speed the super-twisting method shows a good performance. Also, the overshoot values and changes in the control signal range are acceptable in this method, and it was able to control and eliminate the chattering phenomenon. The MSE value is a bit high according to its fluctuations in initial times. In Figures 9.12 and 9.13 only the function $u = -k$ sign(s) cannot effectively track the desired input signal due to the excessive chattering phenomenon. The undesired chattering phenomenon is clearly shown in zoom-in window.

By removing the chattering phenomenon, the tracking error is improved and the reaching time is reduced, which makes the controller more resistant to uncertainties and disturbances.

9.3 MATLAB Programming

The simulation of the above system by the Simulink section in MATLAB software is shown in Figure 9.14. Also, the codes related to the functions are presented below.

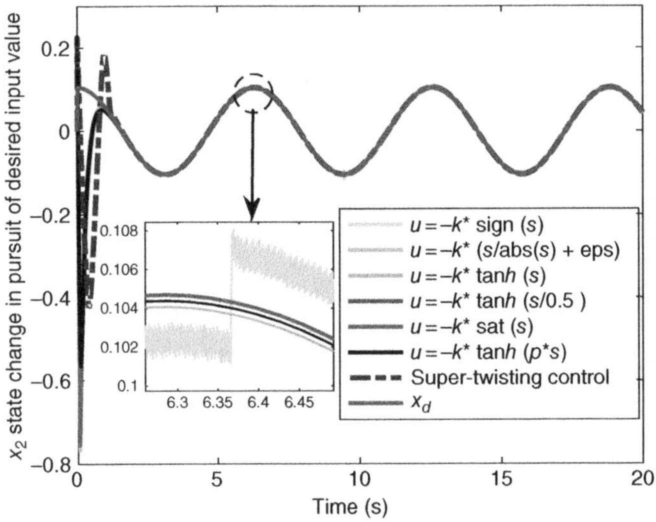

Figure 9.13 Reference tracking by x_2, using controlling structures of Table 9.1.

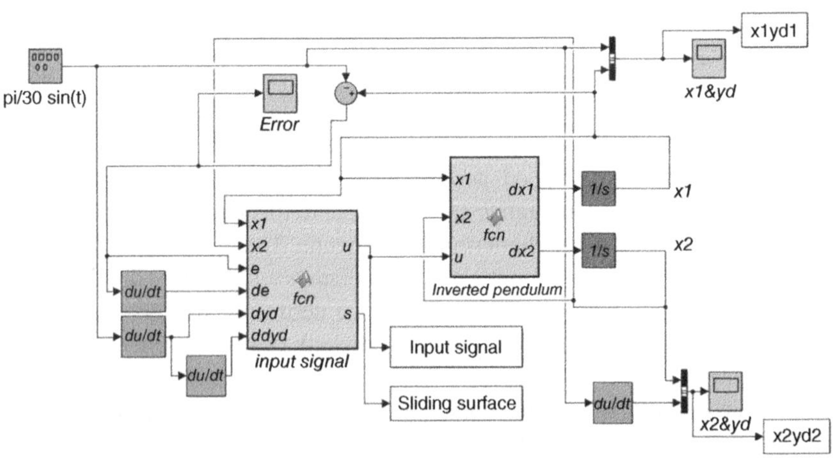

Figure 9.14 Block diagram of the system in Simulink.

The inverted pendulum dynamics is defined in the following MATLAB function:

```
function [dx1,dx2]= fcn(x1,x2,u)
m = 0.1;

M = 1;

L = 0.5;

g = 9.8;

zeros(2,1);

dx1 = x2;

f = ((M+m)*g*sin(x1)-
m*L*(x2^2)*sin(x1)*cos(x1))/((4/3)*(M+m)*L-
m*L*(0.5+0.5*cos(2*x1)));

b = cos(x1)/((4/3)*(M+m)*L-m*L*(0.5+0.5*cos(2*x1)));

dx2 = f+b*u;
```

The input signal function is as follows, where the functions can be activated one by one by removing the "%" behind each function:

```
function [u,s] = fcn(x1,x2,e,de,dyd,ddyd)
m = 0.1;M = 1;L = 0.5;g = 9.8;k = 18;EPS = 0.01;eta = 0.1;
y = 5;p = 1;

s = de + y*e;            %sliding surface
u1 = -k*sign(s);           %Sign function

%u1 = -k*s/(abs(s)+EPS);   %ep-sign function
%u1 = -k*tanh(s);          %Hyperbolic tangent function
%u1 = -k*tanh(s/0.5);      %Generalized hyperbolic tangent
                           function
%u1 = -k*(exp(p*s)-exp(-p*s))/exp(p*s)+exp(-p*s);
%Generalized hyperbolic
% u1 = -k*SAT(s);          %Saturation function
% u1 = -2*(abs(s))^0.5*sign(s)+w;                 % Super
twisting control
```

```
% wdot = -8*sign(s);

f = ((M+m)*g*sin(x1)-m*L*(x2^2)*sin(x1)*cos(x1))/((4/3)*
(M+m)*L-m*L*(0.5+0.5*cos(2*x1)));

b = cos(x1)/((4/3)*(M+m)*L-m*L*(0.5+0.5*cos(2*x1)));

h = y*(x2-dyd)+(f-ddyd);
u = (-1/b)*(h-u1);
end
% function Z = SAT(S)        %Saturation function
% delta = 0.1
% if S>delta
%       Z = 1;
% elseif S
%       Z = -1;
% else
%       Z = 1/delta*S;
% end
% end
```

References

1 Razzaghian, A., Moghaddam, R.K., and Pariz, N. (2021). Fractional-order nonsingular terminal sliding mode control via a disturbance observer for a class of nonlinear systems with mismatched disturbances. *Journal of Vibration and Control* 27 (1–2): 140–151.

2 Razzaghian, A., Kardehi Moghaddam, R., and Pariz, N. (2022). Adaptive fuzzy fractional-order fast terminal sliding mode control for a class of uncertain nonlinear systems. *International Journal of Industrial Electronics, Control and Optimization* 5 (1): 77–87.

3 Qiu, Z.-c. and Zhang, S.-m. (2016). Fuzzy fast terminal sliding mode vibration control of a two-connected flexible plate using laser sensors. *Journal of Sound and Vibration* 380: 51–77.

4 Serdar, C. (2020). Non-linear control of inverted pendulum. *Journal of the Faculty of Engineering and Architecture* 35 (1): 27–38.

5 Ma, Z. and Yu, Q (2022). Control strategy for boost converter based on adaptive integral terminal sliding-mode control. *Journal of Physics: Conference Series, ICPEET-2022.*

6 Liu, X., Xu, X., Zhu, Z., and Jiang, Y. (2021). Dual-arm coordinated control strategy based on modified sliding mode impedance controller. *Sensors* 21: 4653.

7 Zare, A.R. and Ahmadizadeh, M. (2018). Modified sliding mode design of passive viscous fluid control systems for nonlinear structures. *Engineering Structures* 162: 245–256.

8 Jedda, O., Ghabi, J., and Douik, A. (2017). *Sliding Mode Control of an Inverted Pendulum*, Chapter 6. ResearchGate.

9 Saravanakumar, R. and Jena, D. (2015). Validation of an integral sliding mode control for optimal control of a three blade variable speed variable pitch wind turbine. *Electrical Power and Energy Systems* 69: 421–429.

10 Majout, B., Bossoufi, B., Bouderbala, M. et al. (2022). Improvement of PMSG-based wind energy conversion system using developed sliding mode control. *Energies* 15: 1625.

11 Liang, Y., Dong, Z., Li, G., and Wu, T. (2022). Adaptive chattering-free PID sliding mode control for tracking problem of uncertain dynamical systems. *Electronics* 11: 3499.

12 Ansarifar, G.R. and Akhavan, H.R. (2015). Sliding mode control design for a PWR nuclear reactor using sliding mode observer during load following operation. *Annals of Nuclear Energy* 611–619.

13 Ali, S., Prado, A., and Pervaiz, M. (2023). Hybrid backstepping-super twisting algorithm for robust speed control of a three-phase induction motor. *Electronics* 12: 681.

10

Feedback Linearization and Sliding Mode Controllers

10.1 Introduction

In this chapter, the feedback linearization control technique is introduced and implemented on the inverted pendulum nonlinear system. Feedback linearization control technique is a nonlinear controlling methodology that converts nonlinear systems to a linear structure, using a linear virtual controller. In recent years, this technique has effectively applied on many industrial applications including Feedback linearization of nonlinear chaotic systems [1], parametric uncertainty handling of under-actuated nonlinear systems using online optimal input–output feedback linearization controller [2], optimal robust control based on approximate feedback linearization [3], and feedback linearization controller for a distributed solar collector field [4, 5]. Mathematical relations and implementation of the feedback linearization control method are simple compared to similar nonlinear structures. Brockett is the inventor of feedback linearization method for nonlinear control systems. He introduced the concept of feedback linearization in 1978 [6]. In recent years, this method and its developed models have played a significant role in the field of nonlinear systems control. In this chapter, Section 10.1 introduces the feedback linearization method, its applications, and history. In Section 10.2, the feedback linearization method is introduced, and in Section 10.3, the inverted pendulum system is controlled using this method. Further, in Sections 10.4 and 10.5, industrial system controller design and the development of feedback linearization method are done. In Section 10.5, proof of stability and MATLAB programming and implementation of feedback linearization method in Simulink and simulation results are discussed. Finally a collaborative controller including SMC and feedback linearization method is designed in Section 10.6.

Methods of Developing Sliding Mode Controllers: Design and MATLAB Simulation, First Edition.
Reihaneh Kardehi Moghaddam and Mostafa Rabbani.

Published 2025 by John Wiley & Sons, Inc.

10.2 Feedback Linearization

In this section, general concepts of feedback linearization have been introduced. In the feedback linearization method, using a nonlinear mapping of state variables which eliminate the nonlinear terms, the form of state equations governing the system is attempted to be virtually linear. Then, using linear systems controlling methods, a suitable controller is designed. Finally, the real controller is obtained based on the linearized system's virtual controller. It is crucial to note that upon applying the controller, an assessment of the nonlinear system's internal stability is necessary. Investigating internal stability and zero dynamics of the system serves to ensure that unobserved states in the system output do not lead to instability. This analysis helps in understanding the effect of these dynamics on the overall stability of the system. In general, three forms are introduced to display the dynamic equations of the system in this chapter. These three forms are as follows [5]:

1) Feedback linearization and the canonical form
2) Input-state linearization
3) Input–output linearization

10.2.1 Feedback Linearization of Systems with Canonical Form

If the dynamic equations of the system are in canonical form (10.1), with the help of feedback linearization method, nonlinear terms are removed, so that the closed-loop dynamics of the system are linear.

$$\begin{cases} \dot{x}_1 = x_2 \\ \dot{x}_2 = x_3 \\ . \\ . \\ . \\ \dot{x}_n = f(x) + g(x)u(t) \end{cases} \tag{10.1}$$

Here $X = [x_1, x_2, \ldots]^T$ is a vector of state variables. Here a special class of controllable canonical systems is considered. The control input is considered as below (assuming g to be nonzero) [5]:

$$u = \frac{1}{g}(-f + v) \tag{10.2}$$

Replacing Eq. (10.2) in (10.1), the nonlinear terms are removed and the behavior of the system becomes linear in the form of Eq. (10.3).

$$x^n = \nu \tag{10.3}$$

v is a virtual controller and its suggested structure is as follows (10.4):

$$v = x_d^n - k_{n-1}e^{n-1} \cdots k_1\dot{e} - k_0 e \Rightarrow e^n + k_{n-1}\, e^{n-1} + \cdots k_1\dot{e} - k_0 e = 0 \tag{10.4}$$

The stabilizer coefficients k_i are determined in a manner that guarantees all the roots of Eq. (10.4) have a negative real value. In Eq. (10.4), x_d is the desired input signal and $e = x - x_d$ is the tracking error (10.5).

$$\lim_{t \to \infty} \|x - x_d\| = 0 \tag{10.5}$$

10.2.2 Input-State Linearization

In this section, the generalities of the input-state linearization method are briefly stated, and the details of this method are fully described in References [5] and [6].

Consider the following nonlinear system:

$$\dot{x} = f(x, u) \tag{10.6}$$

In this case, first the nonlinear system should be linearized using the nonlinear state mapping $Z = W(X)$, and then the virtual controller v should be designed for the system.

$$\dot{z} = Az + bv \tag{10.7}$$

As system (10.7) is virtually linear, a linear controller is designed for it. The block diagram of the input-state linearization method is shown in Figure 10.1.

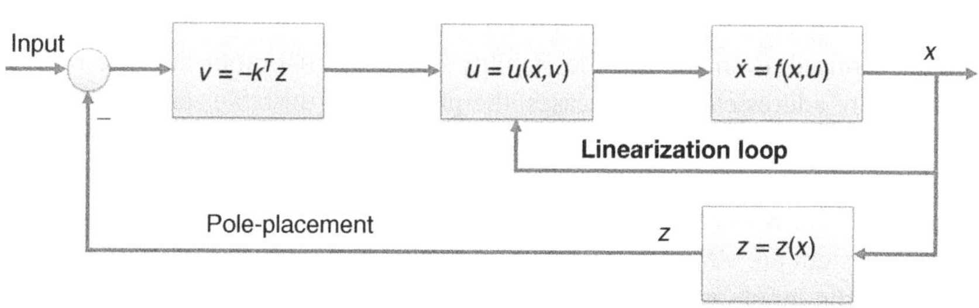

Figure 10.1 Input-state linearization.

10.2.3 Input–Output Linearization

In this section, the generalities of the input–output linearization method are briefly stated, and the details of this method are fully described in References [5] and [6]. Suppose the form of the dynamic equations of the nonlinear system is as follows:

$$\dot{x} = f(x, u) \tag{10.8}$$

$$y = h(x) \tag{10.9}$$

It may be necessary to linearize the input and output response of the system instead of the system dynamics which were linearized in previous methods. Equation (10.8) is a type of tracking problem where the output y is supposed to track the desired input signal y_d in such a way that all states of the system are bounded. It is also assumed that $y_d(t)$ and all its higher derivatives exist and are bounded [5]. In Eq. (10.9), the signal y is indirectly related to u through the changes of state x and nonlinear state Eq. (10.8). To solve this issue, the derivative from output to u must appear in the equations. Then it is calculated with the help of a virtual controller similar to (10.2), and the nonlinear terms are removed. Now the virtual controller can be designed. Since the problem is of tracking type, the design of the virtual controller is in the form of an Eq. (10.4) in terms of error. One of the most important issues that must be investigated in the input–output linearization method is the internal stability. The number of derivatives that must be taken from the output so that the input appears in the output's derivative is called the relative degree of the system. Always the relative degree of the system is less than or equal to the order of the system. For example, if the relative degree of the system is $R = 2$ and the order of the system is $N = 4$, then only 2 system states affect the output. The two states of the system that do not affect the output are called the internal dynamics of the system. Since the internal dynamics is unobservable from the output point of view, the validity of the controller and the possibility of implementing the input–output mode feedback controller exist when the internal dynamics are stable. If the internal dynamics are unstable, the overall stability of the closed-loop system is compromised, even though the stability of the input–output and the tracing problem are addressed. In such cases, the presence of unstable states within the system negatively affects its performance. For example, it may be dangerous for the internal states of the system to grow without limit. In other words, if the primary system is minimum phase or for nonlinear systems, if the zeros of the primary system are stable, it leads to the stabilization of the internal dynamics. Finally, if the input–output feedback linearization method is used to design the controller, the internal stability must be checked to make sure that the states that are not seen in the output do not cause system instability.

10.3 Inverted Pendulum System Equations

The model and parameters of the nonlinear inverted pendulum system are fully described in Chapter 7. Suppose the dynamic model of the system is as follows. The dynamic equations of the system are in the following form:

$$x^n = f(x) + g(x)u \tag{10.10}$$

where the u is input control signal, x are state variables, and $f(x)$ and $g(x)$ are nonlinear functions. The equations of the inverted pendulum system are in the form of Eq. (10.11).

$$\begin{cases} \dot{x}_1 = x_2 \\ \\ \dot{x}_2 = f(x_1, x_2) + g(x_1, x_2)u \end{cases} \qquad \dot{x} = \begin{bmatrix} \theta \\ \dot{\theta} \end{bmatrix} \tag{10.11}$$

The goal is to control the angle of the pendulum so that it follows the desired input reference signal and the error converges to zero. The nonlinear functions f and g expressed in Eq. (10.11) are in the form of Eq. (10.12).

$$\begin{cases} f(x_1, x_2) = \dfrac{(M+m)g\sin x_1 - mLx_2^2(\sin x_1 \cos x_1)}{\frac{4}{3}(M+m)L - mL\cos^2 x_1} \\ \\ g(x_1, x_2) = \dfrac{\cos x_1}{\frac{4}{3}(M+m)L - mL\cos^2 x_1} \end{cases} \tag{10.12}$$

Figure 10.2 shows the limit cycle diagram of the dynamics of the inverted pendulum nonlinear system, or the phase portrait of the system. In this figure, the release point of the system from the initial conditions $x_0 = [0.2, 0.1]$ is marked with a blue star, and the stopping point of the system is marked with a yellow star. As can be seen from Figure 10.2, despite the difference in the initial conditions of the system states, the states converge to an oscillatory cycle.

Also, the nonlinear behavior of state variables x_1 and x_2 is shown in Figures 10.3 and 10.4.

10.4 Controller Design Based on Feedback Linearization

In this chapter, the feedback linearization strategy is used to linearize the nonlinear equations of the inverted pendulum. Considering the canonical form, it is possible to linearize the nonlinear system by using a quasi-linear controller and

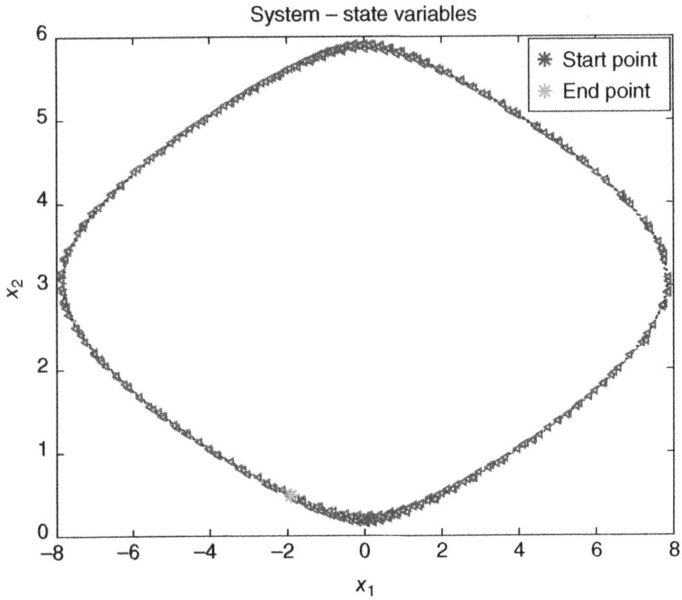

Figure 10.2 The limit cycle of the dynamics of the nonlinear inverted pendulum system.

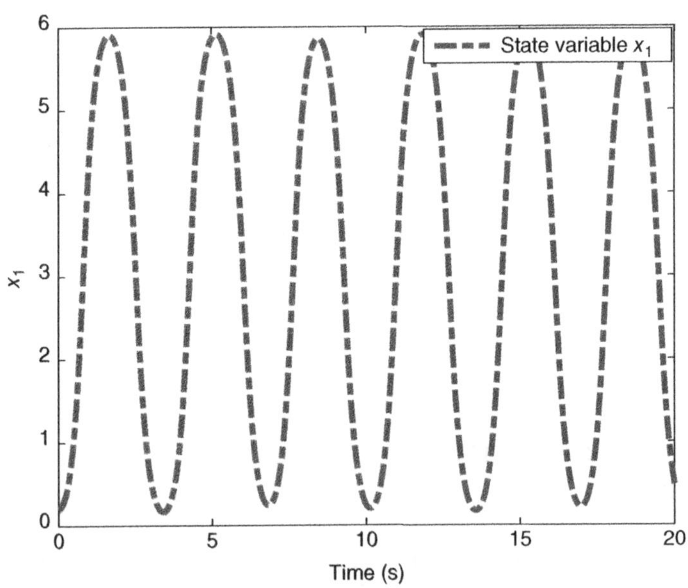

Figure 10.3 Nonlinear behavior of the x_1 state variable.

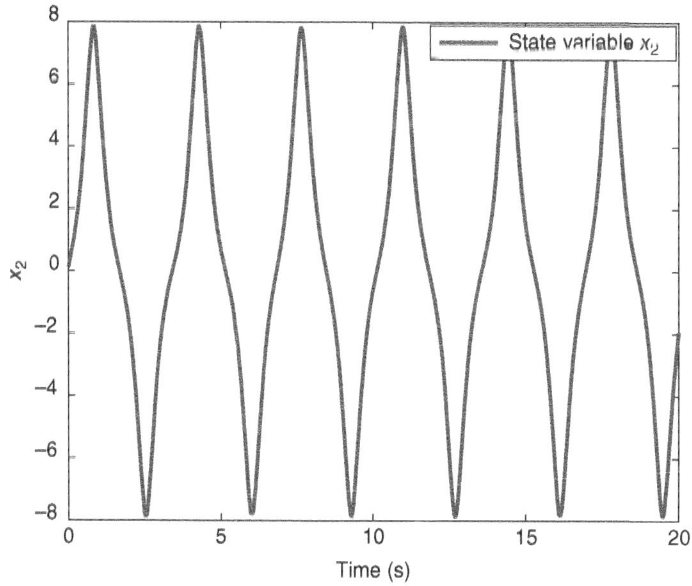

Figure 10.4 Nonlinear behavior of the x_2 state variable.

guarantee the stability of system output variables. According to the canonical structure of system-state equations, the controller signal is proposed as Eq. (10.13).

$$u = \frac{1}{b}(-f + v) \Rightarrow u = \frac{\frac{4}{3}(M + m)L - mL\cos^2 x_1}{\cos x_1} \left(-\frac{(M + m)g\sin x_1 - mLx_2^2(\sin x_1 \cos x_1)}{\frac{4}{3}(M + m)L - mL\cos^2 x_1} + v \right) \tag{10.13}$$

$$\rightarrow \dot{x}_2 = v \tag{10.14}$$

Consequently the nonlinear system became a linear system. For the linear system, the virtual controller is designed as follows:

$$v = \ddot{x}_d - k_1\dot{e} - k_0 e \Rightarrow \ddot{e} + k_1\dot{e} + k_0 e = 0 \tag{10.15}$$

$$s^2 + k_1 s + k_0 = 0 \tag{10.16}$$

We rewrite Eq. (10.15) as (10.15). The coefficients k_1 and k_0 determine the location of the closed-loop poles of the system. Finally, the virtual controller is determined as Eq. (10.16).

$$v = -k_1 x_1 - k_2 x_2 \tag{10.17}$$

The above equation is a linear virtual controller that is used for a nonlinear system with the feedback linearization method. Now, any linear controller can be used to control the nonlinear system, because a linear equivalent model was obtained for the original nonlinear model. Therefore, any linear control method may be used to obtain virtual controller.

10.5 Proof of Stability

In this section, the controller designed in Section 10.4 is implemented on the inverted pendulum system. Accordingly we suppose $x_0 = [0.2,\ 0.1]$ and $x_d = \frac{\pi}{30}\sin(\mathrm{t})$. x_0 is the initial condition and x_d is the input reference signal. To guarantee stability, k_1 and k_0 should be determined in a way that guarantees the Hurwitz property of polynomial (10.16). Accordingly we suppose $k_1 = k_2 = 10$. The controlling signal extracted from feedback linearization method and systems states are illustrated in Figures 10.5 and 10.6, respectively.

10.5.1 MATLAB Programming

The simulation of the above system by MATLAB Simulink is shown in Figure 10.7.

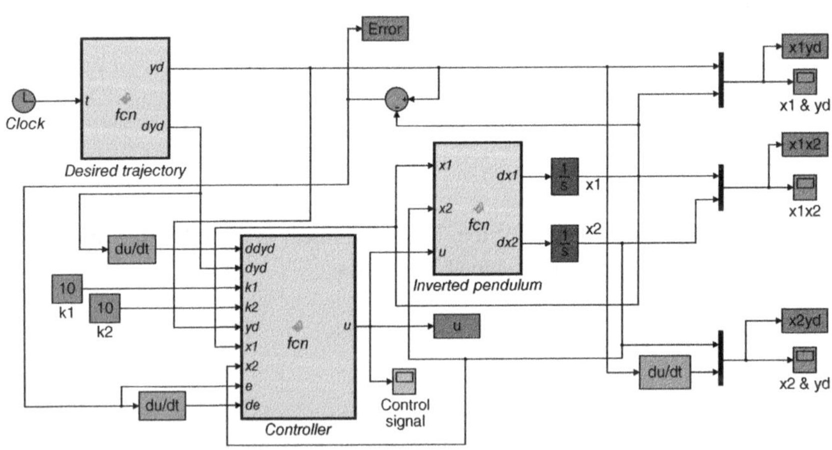

Figure 10.5 State variables x_1 and x_2.

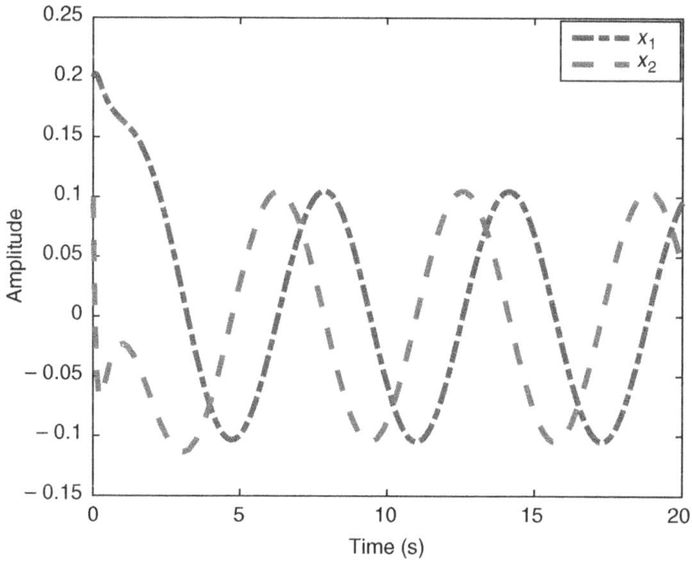

Figure 10.6 Controlling signal.

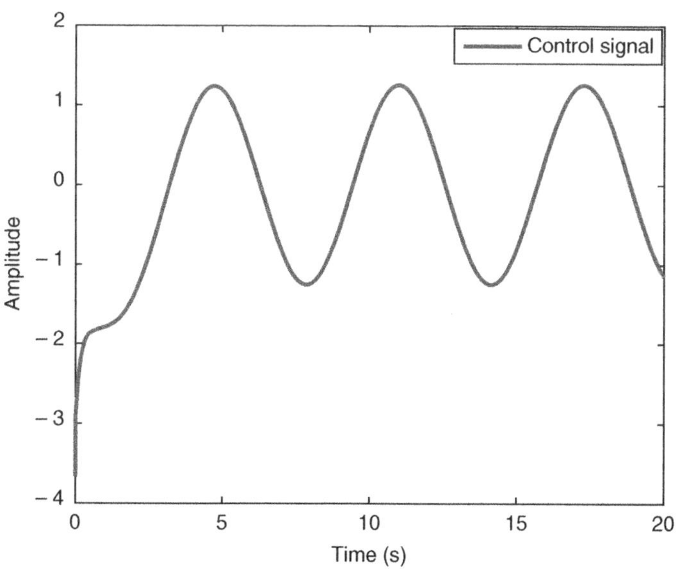

Figure 10.7 Block diagram of the system in Simulink.

The desired trajectory is described as follows:

```
function [yd,dyd] = fcn(t)
yd = (pi/30)*sin(t);
dyd = (pi/30)*cos(t);
```

The inverted pendulum function is as follows:

```
function [dx1,dx2] = fcn(x1,x2,u)
m = 0.1;
M = 1;
L = 0.5;
g = 9.8;
zeros(2,1);
dx1 = x2;
b = cos(x1)/((4/3)*(M+m)*L-m*L*(L+sin(2*x1)));
f = ((M+m)*g*sin(x1)-m*L*(x2^2)*sin(x1)*cos(x1))/((4/3)*
(M+m)*L-m*L*(1+sin(2*x1)));

dx2 = f+b*u;
```

The control function is defined as follows:

```
function u = fcn(ddyd,dyd,k1,k2,yd,x1,x2,e,de)
v = ddyd+k1*de+k2*e;
m = 0.1;
M = 1;
L = 0.5;
g = 9.8;
b = cos(x1)/((4/3)*(M+m)*L-m*L*(L+sin(2*x1)));
f = ((M+m)*g*sin(x1)- m*L*(x2^2)*sin(x1)*cos(x1))/((4/3)*
(M+m)*L-m*L*(1+sin(2*x1)));
u = (1/b)*(-f+v);
```

The error signal is shown in Figure 10.8. Also the tracking error related to states x_1 and x_2 is shown in Figures 10.9 and 10.10 as it is mentioned the error converges to zero in about 5 seconds.

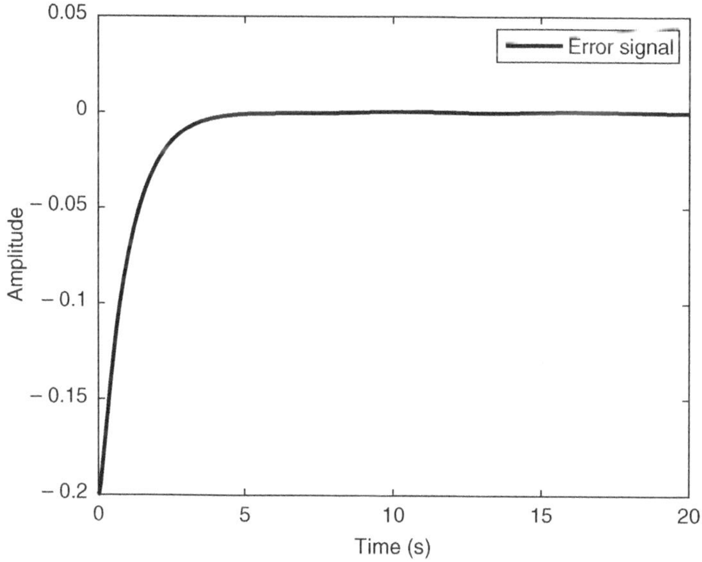

Figure 10.8 Error signal obtained by feedback linearization method.

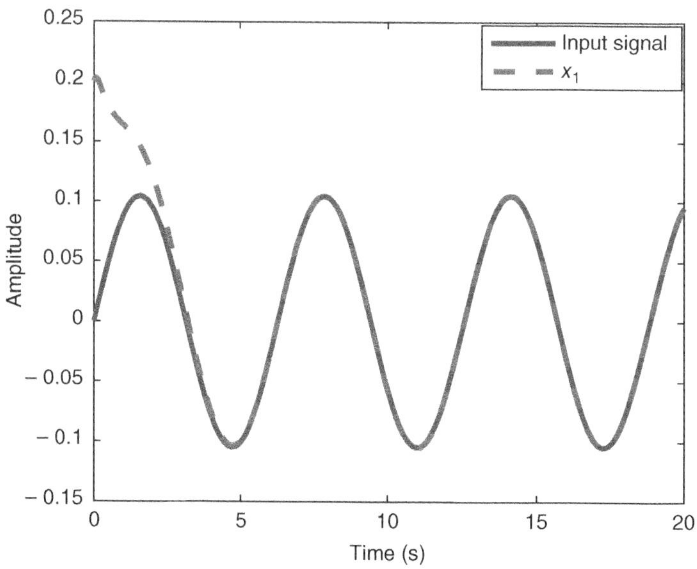

Figure 10.9 x_1 reference tracking error obtained by feedback linearization method.

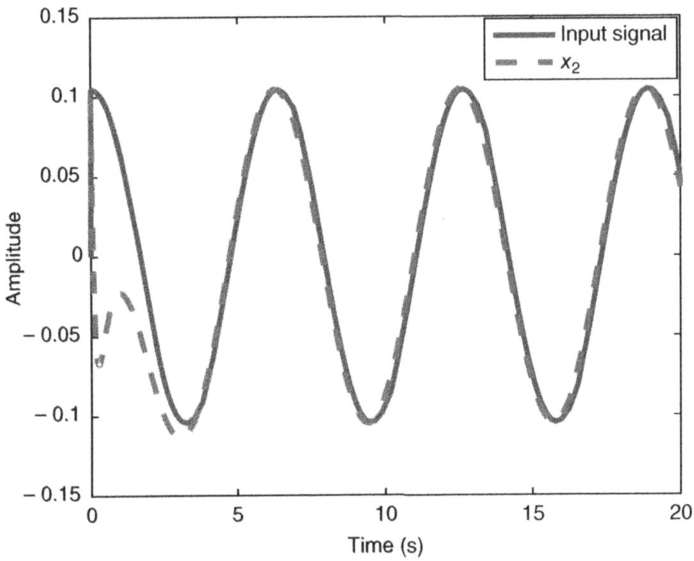

Figure 10.10 x_2 reference tracking error obtained by feedback linearization method.

10.6 Sliding Mode Control Based on Feedback Linearization Method

This section presents a robust nonlinear controller based on the sliding mode and feedback linearization technique. The feedback linearization method is susceptible to uncertainties in system parameters, which can potentially deteriorate the control performance. In cases where the level of uncertainty is high, integrating an adaptive control scheme with feedback linearization may help improve the system response [7, 8]. In the SMC method, the nonlinear terms of the system are first eliminated by the feedback linearization technique [9], and then to improve the system performance in reducing the error, increasing the speed, and making the system resistant to parametric uncertainties, a sliding mode controller is proposed.

Consider the system as follows:

$$\ddot{x} = f(x,t) + g(x,t)u \tag{10.18}$$

$f(x)$ and $g(x)$ are nonlinear functions and u is the control signal. By selecting the control signal in the form of (10.19), nonlinear parts are removed.

$$u = \frac{v - f(x,t)}{g(x,t)} \tag{10.19}$$

In Eq. (10.19), the parameter v is the virtual controller.

Substituting Eq. (10.19) in Eq. (10.18), the dynamics of the system becomes as follows [10]:

$$\ddot{x} = \upsilon \tag{10.20}$$

The sliding surface is selected according to the error as follows:

$$s = \dot{e} + \lambda e \tag{10.21}$$

$$e = x - x_d \tag{10.22}$$

$$\dot{s} = \ddot{e} + \lambda\dot{e} = \ddot{x} - \ddot{x}_d + \lambda\dot{e} = 0 \tag{10.23}$$

$$\ddot{x} = \ddot{x}_d - \lambda\dot{e} \tag{10.24}$$

According to Eq. (10.24), the virtual control was designed as follows:

$$\upsilon = \ddot{x}_d - \lambda\dot{e} \tag{10.25}$$

Now, with the help of sliding mode control, the control signal is defined as follows:

$$u = \hat{u} - u_{SW}; \quad u_{SW} = \frac{k}{g}\,\text{sign}(s) \tag{10.26}$$

$$\hat{u} = \frac{1}{g}(-f + \upsilon) = \frac{1}{g}(-f + \ddot{x}_d - \lambda\dot{e}) \tag{10.27}$$

$$u = \hat{u} - \frac{k}{\hat{g}}\,\text{sign}(s) \Rightarrow u = \frac{1}{\hat{g}}\left(-\hat{f} + \ddot{x}_d - \lambda\dot{e} - k\,\text{sign}(s)\right) \tag{10.28}$$

10.6.1 MATLAB Programming

The simulation of the feedback linearization SMC method in MATLAB Simulink shown in Figure 10.11. In this section, the controller designed in Section 10.6 is applied on the inverted pendulum system. Accordingly $x_0 = [0.2,\ 0]$ and $y_d = \frac{\pi}{30}\sin(t)$. x_0 is the initial condition and y_d is the input reference signal. In the simulations, $K = \lambda = 10$ is chosen.

Figure 10.12 shows the controlling signal which suffers from chattering. The desired trajectory function program is as follows:

```
function [yd,dyd] = fcn(t)
yd = (pi/30)*sin(t);
dyd = (pi/30)*cos(t);
```

The inverted pendulum function program is as follows:

```
function [dx1,dx2] = fcn(x1,x2,u)
m = 0.1;
M = 1;
```

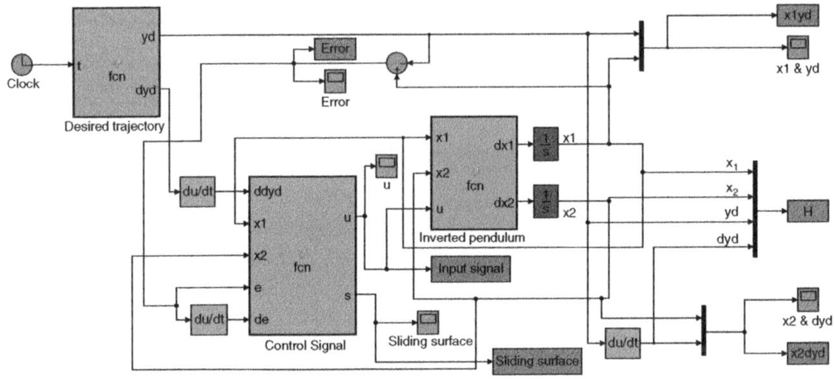

Figure 10.11 Block diagram of the system in Simulink.

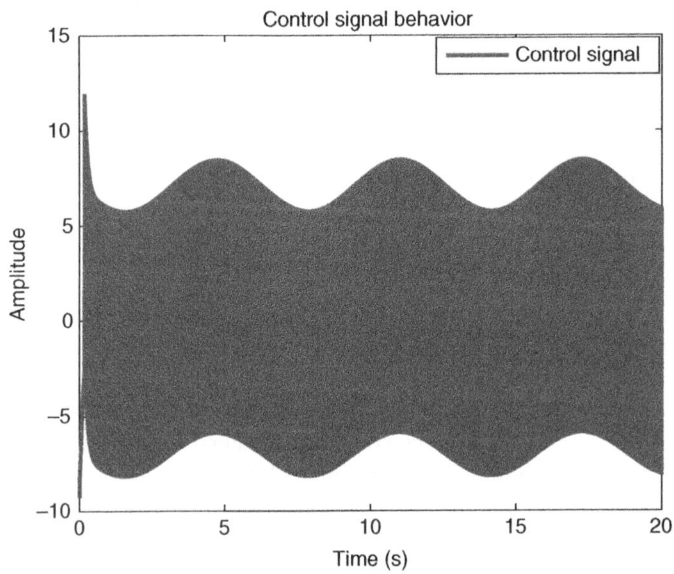

Figure 10.12 Controlling signal.

```
L = 0.5;
g = 9.8;
zeros(2,1);
dx1 = x2;
b = cos(x1)/((4/3)*(M+m)*L-m*L*(L+sin(2*x1)));
f = ((M+m)*g*sin(x1)-m*L*(x2^2)*sin(x1)*cos(x1))/((4/3)*
(M+m)*L-m*L*(1+sin(2*x1)));
dx2 = f + b * u;
```

The control signal function is defined as follows:

```
function [u,s] = fcn(ddyd,x1,x2,e,de)

landa = 10;
k = 10;

s = de + landa * e;

v = ddyd - landa * de;

m = 0.1;
M = 1;
L = 0.5;
g = 9.8;

b = cos(x1)/((4/3)*(M+m)*L-m*L*(L+sin(2*x1)));

f = ((M+m)*g*sin(x1)-m*L*(x2^2)*sin(x1)*cos(x1))/((4/3)*
(M+m)*L-m*L*(1+sin(2*x1)));

u = (1/b)*(-f + v - k * sign(s));
```

The control signal shown in Figure 10.13 suffers from chattering. To handle this issue, some enhancement methods introduced in Chapter 9 are considered. Figure 10.13 shows the sliding surface, which has fluctuations.

As seen in Figure 10.14, in the feedback linearization SMC method, the error converges to zero in less than 2 seconds as illustrated in Figures 10.15 and 10.16.

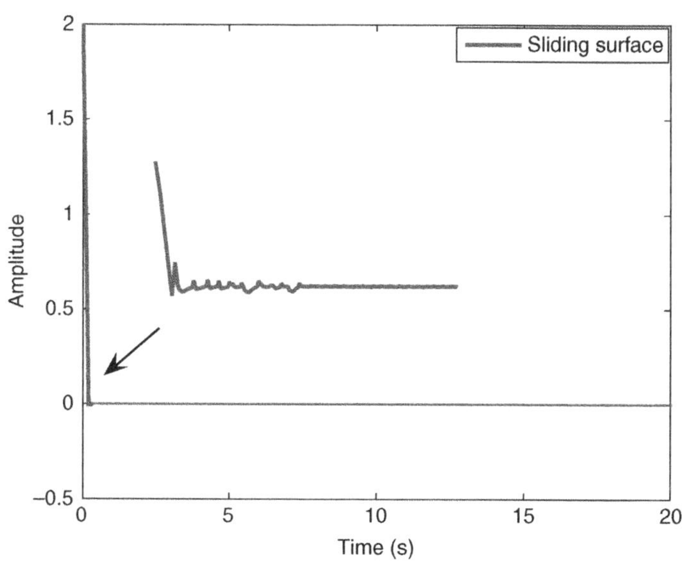

Figure 10.13 Sliding surface in feedback linearization SMC method.

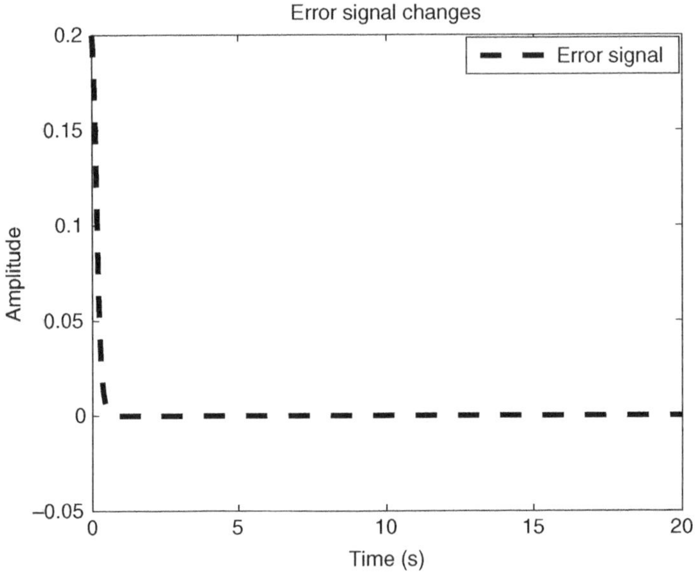

Figure 10.14 Error signal obtained by feedback linearization SMC method.

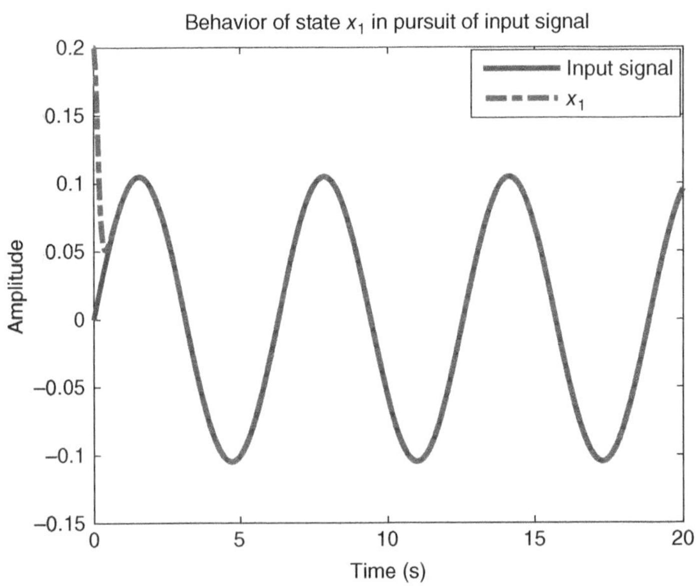

Figure 10.15 x_1 reference tracking error obtained by feedback linearization SMC method.

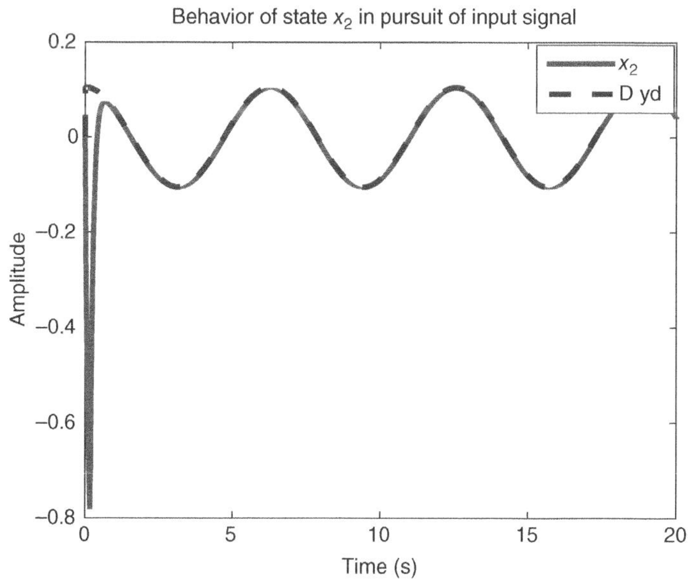

Figure 10.16 x_2 reference tracking error obtained by feedback linearization SMC method.

References

1 Kabanov, A.A. (2020). Feedback linearization of nonlinear singularly perturbed systems with state-dependent coefficients. *International Journal of Control, Automation and Systems* 18: 1–8.

2 Mahmoodabadi, M.J. and Andalib Sahnehsaraei, M. (2021). Parametric uncertainty handling of underactuated nonlinear systems using an online optimal input–output feedback linearization controller. *Systems Science and Control Engineering* 9: 209–218.

3 Sahnehsaraei, M.A. and Mahmoodabadi, M.J. (2020). Approximate feedback linearization based optimal robust control for an inverted pendulum system with time-varying uncertainties. *International Journal of Dynamics and Control* 9: 160–172.

4 Cirrea, C.M., Berenguelb, M., Valenzuelac, L., and Camacho, E.F. (2007). Feedback linearization control for a distributed solar collector field. *Control Engineering Practice* 15: 1533–1544.

5 Slotine, J. and Li, W. (1990). *Applied Nonlinear Control*. Prentice Hall.

6 Krener, A.J. (1999). *Mathematical Control Theory*. Springer.

7 Li, P., Xiong, L., Fei, W. et al. (2019). Sliding mode controller based on feedback linearization for damping of subsynchronous control interaction in DFIG-based wind power plants. *Electrical Power and Energy Systems* 107: 239–250.

8 Daewon Lee, H., Kim, J., and Sastry, S. (2009). Feedback linearization vs. adaptive sliding mode control for a quadrotor helicopter. *International Journal of Control, Automation and Systems* 7 (3): 419–428.

9 Devanshu, A., Singh, M., and Kumar, N. (2018). Sliding mode control of induction motor drive based on feedback linearization. *IETE Journal of Research* 66: 256–269.

10 Tuan, L.A., Lee, S.-G., Ko, D.H., and Nho, L.C. (2013). Combined control with sliding mode and partial feedback linearization for 3D overhead cranes. *International Journal of Robust and Nonlinear Control*. https://doi.org/10.1002/rnc.3061.

11

Fixed-Time Terminal Sliding Mode Controller

11.1 Introduction

In classic sliding mode control, the state variables reach the sliding surface and converge asymptotically. Although the sliding mode control parameters, introduced in the previous chapters, can be chosen so that the convergence speed increases to a great extent, the system-state variables cannot converge to the equilibrium point in a finite time. In other words, when a controller is designed to ensure the asymptotic stability of the closed-loop system, the convergence of states to the equilibrium point occurs over an infinite time horizon. Meanwhile, in many real-world applications and industrial devices, convergence is required within a finite time. Recently several methodologies have been proposed to guarantee a finite-time convergence [1–3]. Although the finite-time control can stabilize the system in a predetermined time, it is not independent of the initial conditions. This issue declines the effectiveness of the mentioned controlling methods in industrial applications. Fixed-time stability was proposed in 2012 by Andrey Polyakov [4]. This method not only reduces the convergence time but also keeps it independent of the initial conditions of the system. Based on the description, this chapter proposes a fixed-time terminal sliding mode controller for second-order nonlinear systems in the presence of uncertainty and external disturbances. The introduced method brings desirable features such as fast convergence of system-state variables, elimination of singularity, and handle the issue of dependence on initial conditions. In the sequel in Section 11.2 the finite-time terminal sliding mode controller is presented, and the dependency of finite convergence time to initial conditions is investigated; in Section 11.3 the fixed-time terminal sliding mode control is proposed and its characteristics are clearly discussed and compared with finite-time sliding mode controller.

Methods of Developing Sliding Mode Controllers: Design and MATLAB Simulation, First Edition.
Reihaneh Kardehi Moghaddam and Mostafa Rabbani.

Published 2025 by John Wiley & Sons, Inc.

11.2 Finite-Time Terminal Sliding Mode Controller

In this section, the finite-time terminal sliding mode control method for a class of nonlinear systems is analyzed and investigated. In this method, the control signal is designed in such a way that it can guarantee the finite-time convergence of the system's states.

11.2.1 Finite-Time Convergence

Finite-time terminal sliding mode controllers enable system states to reach a neighborhood of equilibrium points within a finite time, though this time is dependent on the initial conditions of the system. In finite-time convergence, the convergence time is directly influenced by their initial conditions; the greater the distance from equilibrium, the longer it takes for the states to converge. So, the convergence of states depends on the initial conditions.

11.2.2 Mathematical Preliminaries

In this section some basic definitions and lemmas are proposed to clarify the stability of proposed controller. Consider the second-order nonlinear system as follows:

$$\begin{cases} \dot{x}_1 = x_2 \\ \dot{x}_2 = f(x,t) + g(x,t)u(t) + d(t) \end{cases} \tag{11.1}$$

In the equation above, $x = [x_1, x_2]^T \in R^n$ is the system-states vector. x_d is the desired trajectory, $\tilde{x}_1 = x_1 - x_d$ is the tracking error, and $\tilde{x}_2 = x_2 - \dot{x}_d$. $f(x, t)$ and $g(x, t)$ are the nonlinear functions, and $u(t)$ is the input control signal. Also, function $d(t) \leq \|D\|$ denotes the bounded disturbance.

Definition 11.1 Finite-Time Stability

The equilibrium point of the nonlinear system defined in (11.1) is finite-time stable if and only if:

- It is stable in the sense of Lyapunov.
- There exists an open neighborhood N_{xe} of the equilibrium point and a certain positive function $T(x_0) : N_{xe} \rightarrow R^+$(settling time) in such a way that it holds the following conditions for all $x_0 \in N_{xe}/\{0\}$.

$$\lim_{t \rightarrow T(x_0)} x(t) = 0 \qquad x(t) = 0, \forall t \geq T(x_0) \tag{11.2}$$

Remark 11.1

The equilibrium point is globally stable if it is finite-time stable for $N_{xe} = R^n$.

Lemma 11.1

For nonlinear system (11.1), suppose that there exists a Lyapunov function $V(x)$ defined on a neighborhood $D \subset R^n$ of the origin. The equilibrium point is locally finite-time stable if the conditions (11.3) are satisfied. In addition, the settling time is considered to satisfy (11.4) [5].

$$\begin{aligned} &1)\ \ V(0) = 0 \\ &2)\ \ V(x) > 0, \forall x \in D/\{0\} \\ &3)\ \ \dot{V}(x) + cV^{\alpha}(x) \le 0, \forall x \in D/\{0\} \end{aligned} \tag{11.3}$$

where $0 < \alpha < 1$ and $c > 0$

Then the origin of system is finite-time stable. The settling time depends on the initial state $x(0) = x_0$, and its upper bound is [6]

$$T(x_0) \le \frac{1}{c(1-\alpha)} v^{1-\alpha}(x_0) \tag{11.4}$$

In the finite-time terminal sliding mode technique [7], sliding surface is defined as follows:

$$s = \tilde{x}_2 + \beta \tilde{x}_1^{\frac{q}{p}} \tag{11.5}$$

In the above equation $\beta > 0$, $p > q$, and p, q are positive odd numbers.

It should be proved that the above sliding surface provides finite-time stability in the sliding phase. This is done as follows. In fact, it will be proven that the modes that reach the sliding surface after a finite time and remain on the surface can converge to the origin in a finite time.

$$s = 0 \Rightarrow \tilde{x}_2 = -\beta \tilde{x}_1^{\frac{q}{p}} \Rightarrow \dot{\tilde{x}}_1 = -\beta \tilde{x}_1^{\frac{q}{p}} \tag{11.6}$$

According to the following relations, the above system has finite-time stability.

$$\lim_{t \to T} \tilde{x}_1(t) = 0$$

$$\lim_{t \to T} \tilde{x}_2(t) = 0$$

Therefore, the finite-time stability of the system states in the sliding phase was guaranteed. According to the finite-time stability theorem, the upper bound of the convergence time is obtained from Eq. (11.4).

11.2.3 Finite-Time Terminal Sliding Mode Controller Design

Theorem 11.1
For system 11.1, u defined below is a finite-time terminal sliding mode controller.

$$u = -g^{-1}\left(f + \beta\frac{q}{p}\tilde{x}_1^{\frac{q}{p}-1}\tilde{x}_2 - \ddot{x}_d + (D+\eta)\text{sign}(s)\right),\ \ \eta > 0 \tag{11.7}$$

where s is the sliding surface, and its derivative is defined as follows:

$$\dot{s} = \dot{\tilde{x}}_2 + \beta\frac{q}{p}\tilde{x}_1^{\frac{q}{p}-1}\dot{\tilde{x}}_1 \tag{11.8}$$

Proof: According to Definition 11.1 and Lemma 11.1, the following conditions must be satisfied:

$$\dot{V} \leq -cV^{\alpha}, c > 0, 0 < \alpha < 1 \tag{11.9.1}$$

Substituting (11.1) in (11.8) leads to (11.9).

$$\dot{s} = f + gu + d - \ddot{x}_d + \beta\frac{q}{p}\tilde{x}_1^{\frac{q}{p}-1}\dot{\tilde{x}}_1 \tag{11.9}$$

The Lyapunov function is proposed as follows:

$$V(s) = \frac{1}{2}s^2 \tag{11.10}$$

The derivative of the Lyapunov function is:

$$\dot{V}(s) = s\dot{s} = s\left(f + gu + d - \ddot{x}_d + \beta\frac{q}{p}\tilde{x}_1^{\frac{q}{p}-1}\tilde{x}_2\right) \tag{11.11}$$

u should be designed in such a way that the finite-time stability condition (11.3) is satisfied.

$$u = -g^{-1}\left(f + \beta\frac{q}{p}\tilde{x}_1^{\frac{q}{p}-1}\tilde{x}_2 - \ddot{x}_d + (D+\eta)\text{sign}(s)\right),\ \ \eta > 0 \tag{11.12}$$

Substituting control law u in Eq. (11.11), $\dot{V}$ is obtained as follows:

$$\dot{V}(s) = sd(t) - (D+\eta)s\,\text{sign}(s) \Rightarrow sd(t) - D|s| - \eta|s| \leq -\eta|s| \tag{11.13}$$

Since $-D|s| + sd \leq 0$, the above relationship is considered. To prove the finite-time stability, (11.3) should be satisfied.

$$V = \frac{1}{2}s^2 \Rightarrow V^{\frac{1}{2}} = \frac{1}{\sqrt{2}}|s| \Rightarrow |s| = \sqrt{2}V^{\frac{1}{2}} \Rightarrow -\eta|s| = -\eta\sqrt{2}V^{\frac{1}{2}}, \gamma \triangleq \eta\sqrt{2}$$

Equation (11.13) is rewritten as follows:

$$\dot{V} \le -\gamma V^{\frac{1}{2}} \tag{11.14}$$

According to the above relation, (11.9) and finite-time stability are guaranteed. □

11.2.4 MATLAB Programming

Consider the inverted pendulum as below:

$$\begin{cases} f(x,t) = \dfrac{(M+m)g\sin x_1 - mLx_2^2(\sin x_1 \cos x_1)}{\frac{4}{3}(M+m)L - mL\cos^2 x_1} \\ g(x,t) = \dfrac{\cos x_1}{\frac{4}{3}(M+m)L - mL\cos^2 x_1} \end{cases}$$

The controller designed in Section 11.2 is applied on the system. Accordingly, we suppose x_0= [0.2, 0.1] and $x_d = \frac{\pi}{30}\sin(t)$. x_0 is the initial condition and x_d is the input reference signal. The disturbance entered into the system is as follows:

$$d(t) = 0.1\sin(t) \tag{11.15}$$

The controlling parameters are considered to be as follows:

$$\mu = 15, \quad D = 0.1, \quad beta = 1, \quad p = 9, \quad q = 3$$

The inverted pendulum function is as follows and the controlling block diagram is illustrated in Figure 11.1

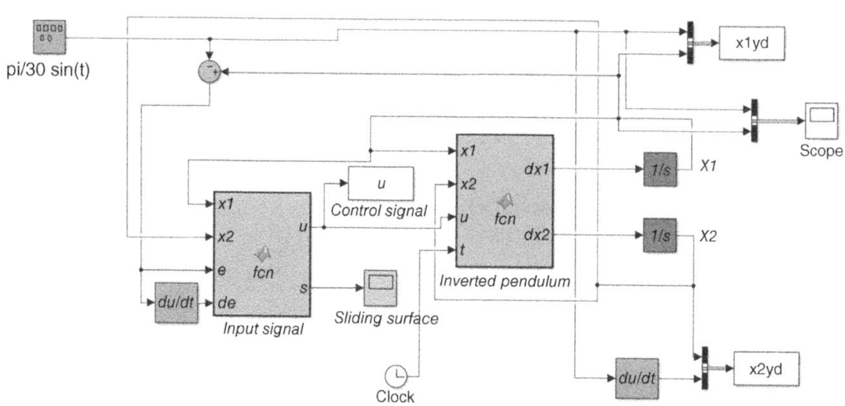

Figure 11.1 Block diagram of finite-time terminal sliding mode control of inverted pendulum.

```
function [dx1,dx2] = fcn(x1,x2,u,t)

m = 0.1;M = 1;L = 0.5;g = 9.8;
zeros(2,1);
dx1 = x2;

f = ((M+m)*g*sin(x1)-m*L*(x2^2)*sin(x1)*cos(x1))/((4/3)*
(M+m)*L-...
    m*L*(0.5+0.5*cos(2*x1)));

g = cos(x1)/((4/3)*(M+m)*L-m*L*(0.5+0.5*cos(2*x1)));

d = 0.1*sin(t);

dx2 = f+g*u+d;
```

The input signal MATLAB function is as follows:

```
function [u,s] = fcn(x1,x2,e,de)

m = 0.1; M = 1; L = 0.5; g = 9.8; eta = 15; beta = 1; p = 9;
q = 3; D = 0.1;

T1 = abs(e)^(q/p)*sign(e);

T2 = abs(x1)^(q/p-1)*sign(x1);

s = de+beta*T1;

f = ((M+m)*g*sin(x1)-m*L*(x2^2)*sin(x1)*cos(x1))/((4/3)*
(M+m)*L...
    -m*L*(0.5+0.5*cos(2*x1)));

g = cos(x1)/((4/3)*(M+m)*L-m*L*(0.5+0.5*cos(2*x1)));

u = (-1/g)*(f+beta*q/p* x2 *T2+(D+eta)*sign(s));

end
```

Figure 11.2 shows the performance of the finite-time terminal sliding mode control method with a low error rate and acceptable accuracy.

Figure 11.3 shows the control signal with bounded fluctuations. Figure 11.4 indicates the Simulink details. To investigate the dependence on the initial conditions

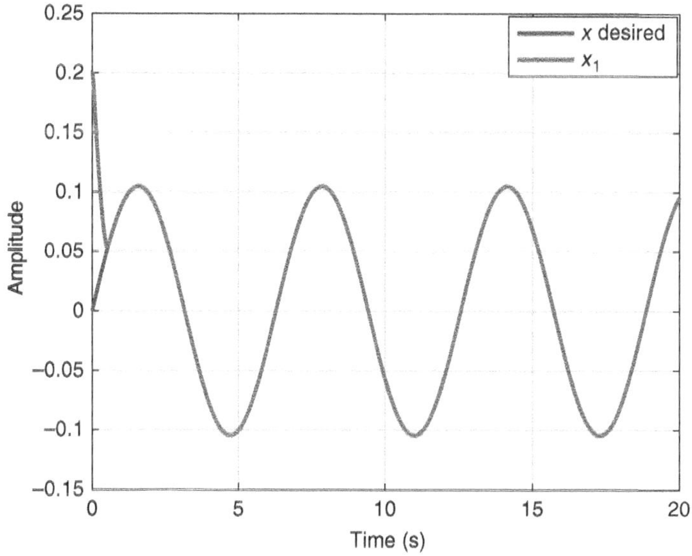

Figure 11.2 Inverted pendulum position tracking using finite-time TSMC.

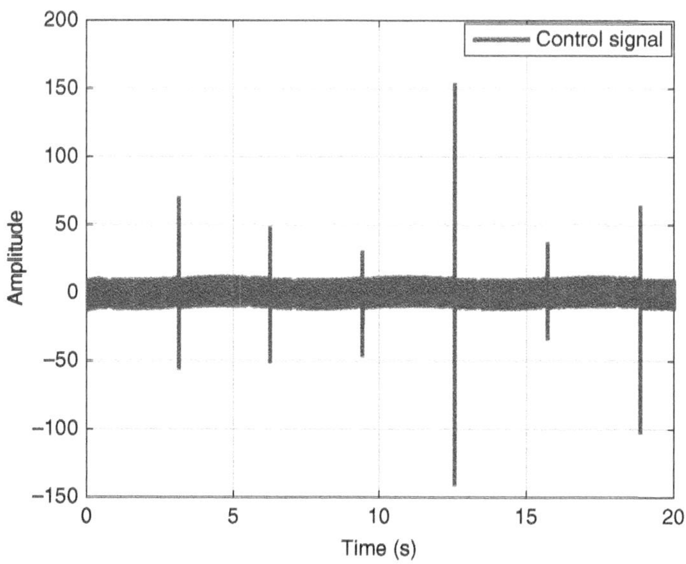

Figure 11.3 Inverted pendulum controller using finite-time TSMC.

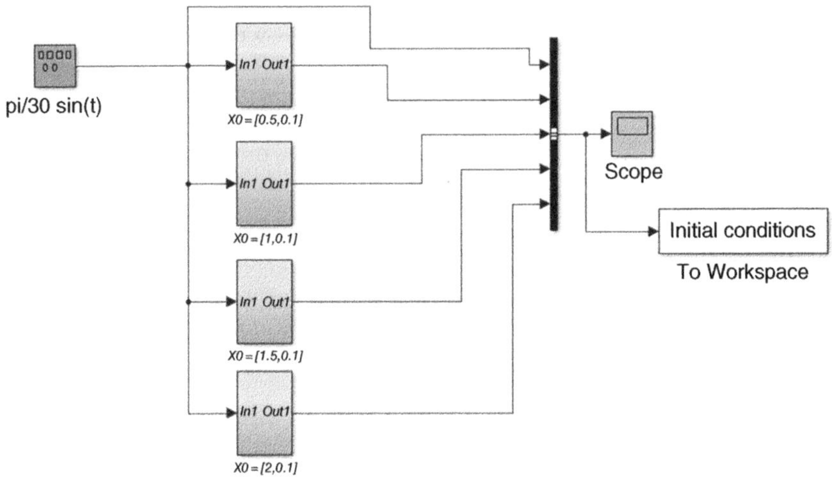

Figure 11.4 Block diagram of the system in Simulink for different initial conditions.

in the finite-time terminal sliding mode control method, system (11.1) has been implemented for different initial conditions. As shown in Figure 11.5, the further the initial conditions are from the equilibrium points of the system, the convergence time increases.

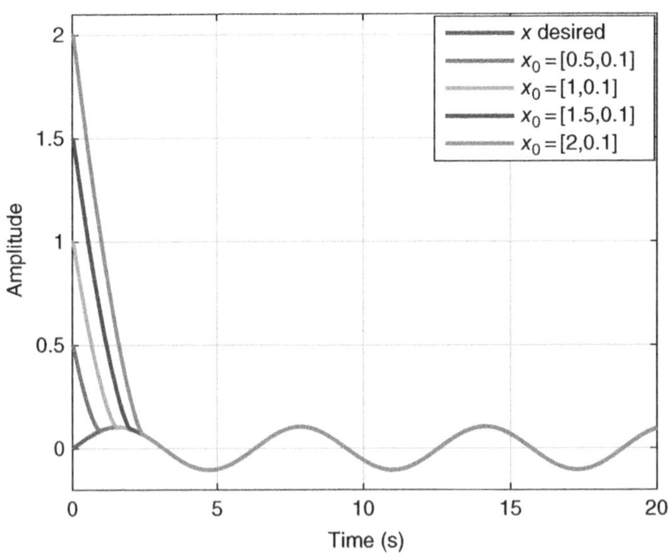

Figure 11.5 Inverted pendulum position tracking using finite-time TSMC for different initial conditions.

11.3 Fixed-Time Terminal Sliding Mode Controller

In this section, the fixed-time terminal sliding mode control method for a class of nonlinear systems defined in (11.1) is analyzed. As demonstrated in Section 11.2, finite-time control can stabilize a system within a predetermined time frame but lacks the ability to guarantee convergence of system trajectories in a finite time, independent of initial conditions. This limitation poses challenges for many industrial applications. To address these issues, a fixed-time stabilization control method is proposed, offering a potential solution for these problems. Considering the mentioned favorable feature, fixed-time controllers are widely used in various industrial systems including multi-agent systems [8] and spacecrafts [9].

11.3.1 Fixed-Time Convergence

In this section a fixed-time terminal sliding mode control is proposed. The sliding structure provides fast convergence and accurate tracking and independence from initial conditions. The region of convergence of system-state variables is presented transparently and based on control parameters. Therefore, it is possible to achieve optimal control accuracy by adjusting the controller parameters.

11.3.2 Mathematical Preliminaries

In this part some basic definitions and lemmas are proposed to clarify the stability of proposed controller.

Definition 11.2 Fixed-Time Stability
The equilibrium point of the nonlinear system, defined in (11.1), is **fixed-time stable** if and only if [10]:

- Its equilibrium point is globally finite-time stable.
- The settling-time $T(x_0)$ is limited according to the initial conditions. In other words, the upper limit of the settling-time T_{max} is independent of the initial conditions, that is $\exists T_{max}$ such that $T(x_0) \leq T_{max}$ for $\forall x_0 \in \mathbb{R}^n$.

Lemma 11.2
Consider the following system.

$$\dot{y} = -\alpha y^{\frac{m}{n}} - \beta y^{\frac{p}{q}}, \quad y(0) = y_0 \tag{11.16}$$

where $\alpha, \beta > 0$, and m, n, q, p are positive odd integers satisfying $0 < n < m$ and $0 < p < q$. Then the equilibrium of (11.16) is fixed-time stable and the settling-time T is bounded as follows:

$$T(x_0) \leq T \frac{1}{\alpha} \frac{n}{m-n} \frac{1}{\beta} \frac{q}{q-p_{max}} \tag{11.17}$$

Remark 11.2

As illustrated in Lemma 11.2, the settling-time function has a limited upper bound which is independent of the initial conditions. In other words, it is only related to the design parameters m, n, q, p, α, and β. Consequently, the time of convergence to the equilibrium points can be guaranteed.

Lemma 11.3

For the nonlinear system (11.16), assume that a continuously differentiable and radially infinite Lyapunov function $V(x): D \rightarrow \mathbb{R}$ exists such that for each $\alpha, \beta >$ 0 the following conditions hold. Then the origin will be fixed-time stable and the upper limit of the convergence time is calculated according to Eq. (11.17) and is independent of the initial conditions.

$$\begin{aligned} &1)\ V(0) = 0 \\ &2)\ V(x) > 0, \quad \forall x \in D \setminus \{0\} \\ &3)\ \dot{V}(x) \leq -\alpha V^{p}(x) - \beta V^{q}(x), \quad \forall x \in D \setminus \{0\} \end{aligned} \tag{11.18}$$

where α, β, p, q, and r are positive constants, satisfying the following conditions.

$$p = 1 - \frac{1}{r}, q = 1 + \frac{1}{r}, r > 1$$

11.3.3 Design and Stability Analysis

To design the fixed-time terminal sliding mode controller the following steps are considered:

- The sliding surface is defined as follows [10–13]:

$$s = \tilde{x}_2 + \alpha_1 \tilde{x}_1^{\frac{m_1}{n_1}} + \beta_1 \tilde{x}_1^{\frac{p_1}{q_1}}, (\alpha_1, \beta_1) > 0 \tag{11.19}$$

  Here m_1, n_1, p_1, q_1 are odd positive integers. Also, the following conditions must be observed.

$$m_1 > n_1 > 0, \quad 0 < p_1 < q_1 < 2p_1$$

- Sliding phase

  In the sliding phase, the following relation is considered:

$$s = 0 \Rightarrow \tilde{x}_2 = -\alpha_1 \tilde{x}_1^{\frac{m_1}{n_1}} - \beta_1 \tilde{x}_1^{\frac{p_1}{q_1}} \Rightarrow \dot{\tilde{x}}_1 = -\alpha_1 \tilde{x}_1^{\frac{m_1}{n_1}} - \beta_1 \tilde{x}_1^{\frac{p_1}{q_1}} \tag{11.20}$$

Remark 11.3

According to Lemma 11.2, system defined in (11.16) is fixed-time stable. In other words, the states converge to the origin in a fixed time.

- Reaching phase

  u should be designed in such a way that states reach the sliding surface in a fixed time. The controller is considered to be $u = \hat{u} - \mu g^{-1} \text{sign}(s)$, where $\hat{u}$ is the nominal part of controller and should guarantee the systems maintenance on sliding surface which is equal to $\dot{s} = 0$.

$$\dot{s} = \dot{\tilde{x}}_2 + \alpha_1 \frac{m_1}{n_1} \tilde{x}_1^{\frac{m_1}{n_1} - 1} \dot{\tilde{x}}_1 + \beta_1 \frac{p_1}{q_1} \tilde{x}_1^{\frac{p_1}{q_1} - 1} \dot{\tilde{x}}_1 \tag{11.21}$$

Substituting (11.1) in (11.21) yields:

$$\dot{s} = f + gu + d - \ddot{x}_d + \alpha_1 \frac{m_1}{n_1} \tilde{x}_1^{\frac{m_1}{n_1} - 1} \tilde{x}_2 + \beta_1 \frac{p_1}{q_1} \tilde{x}_1^{\frac{p_1}{q_1} - 1} \tilde{x}_2 \tag{11.22}$$

The nominal controller is selected as follows:

$$\hat{u} = -g^{-1}\left(f - \ddot{x}_d + \alpha_1 \frac{m_1}{n_1} \tilde{x}_1^{\frac{m_1}{n_1} - 1} \tilde{x}_2 + \beta_1 \frac{p_1}{q_1} \tilde{x}_1^{\frac{p_1}{q_1} - 1} \tilde{x}_2\right) \tag{11.23}$$

Theorem 11.2
For nonlinear system (11.1), the following controller is a stabilizing controller.

$$u = -g^{-1}\left(f - \ddot{x}_d + \alpha_1 \frac{m_1}{n_1} \tilde{x}_1^{\frac{m_1}{n_1} - 1} \tilde{x}_2 + \beta_1 \frac{p_1}{q_1} \tilde{x}_1^{\frac{p_1}{q_1} - 1} \tilde{x}_2 + \alpha_2 s^{\frac{m_2}{n_2}} + \beta_2 s^{\frac{p_2}{q_2}} + (\mu + D)\text{sign}(s)\right) \tag{11.24}$$

$+ \alpha_2 s^{\frac{m_2}{n_2}} + \beta_2 s^{\frac{p_2}{q_2}}$ terms are added to controller to guarantee the stability. The controlling parameters are defined as follows:

$$\alpha_1, \beta_1, \alpha_2, \beta_2 > 0, m_1 > n_1 > 0, 0 < p_1 < q_1 < 2p_1, m_2 > n_2, q_2 > p_2 \tag{11.25}$$

Proof: To prove the stability, the following Lyapunov function is considered:

$$V(s) = \frac{1}{2}s^2 \tag{11.26}$$

Derivation of (11.26) yields:

$$\dot{V}(s) = s\dot{s} = s\left(f + gu + d - \ddot{x}_d + \alpha_1 \frac{m_1}{n_1} \tilde{x}_1^{\frac{m_1}{n_1} - 1} \tilde{x}_2 + \beta_1 \frac{p_1}{q_1} \tilde{x}_1^{\frac{p_1}{q_1} - 1} \tilde{x}_2\right) \tag{11.27}$$

Substituting controller (11.24) in (11.27) yields:

$$\dot{V}(s) = s\dot{s} = s\left(-\alpha_2 s^{\frac{m_2}{n_2}} - \beta_2 s^{\frac{p_2}{q_2}} - (\mu + D)\text{sign}(s) + d\right) \tag{11.28}$$

$$\dot{V}(s) = -\alpha_2 s^{\frac{m_2}{n_2} + 1} - \beta_2 s^{\frac{p_2}{q_2} + 1} - (\mu + D)|s| + sd \le -\alpha_2 s^{\frac{m_2}{n_2} + 1} - \beta_2 s^{\frac{p_2}{q_2} + 1} \tag{11.29}$$

Considering $(\mu + D)|s| \geq sd(t)$, it is concluded that $\dot{V}(s) < 0$ which proves the system stability.

Theorem 11.3
Controller (11.24) is a fixed-time stable controller for system (11.1).

Proof: To prove the fixed-time stability, the following Lyapunov function is considered

$$V(s) = \frac{1}{2}s^2 \rightarrow s = \sqrt{2V}$$

Formula (11.29) is rewritten as below:

$$\dot{V}(s) \leq -\alpha_2\left(\sqrt{2V}\right)^{\frac{m_2}{n_2}+1} - \beta_2\left(\sqrt{2V}\right)^{\frac{p_2}{q_2}+1}$$

y is defined as follows:

$$y \triangleq \sqrt{2V} \rightarrow \dot{y} = \frac{2\dot{V}}{2\sqrt{2V}} \Rightarrow \dot{V} = \sqrt{2V}\dot{y} = y\dot{y}$$

Consequently, it yields:

$$\dot{V} = y\dot{y} \leq -\alpha_2 y^{\frac{m_2}{n_2}+1} - \beta_2 y^{\frac{p_2}{q_2}+1} \rightarrow \dot{y} \leq -\alpha_2 y^{\frac{m_2}{n_2}} - \beta_2 y^{\frac{p_2}{q_2}}$$

According to Lemma 11.3, system is fixed-time stable.

11.3.4 MATLAB Programming

In this section, the controller designed in Section 11.3 is applied on the inverted pendulum system. Accordingly, initial conditions and disturbance are supposed to be $x_0 = [0.2, 0.1]$ and $x_d = \frac{\pi}{30}\sin(t)$. The controlling parameters are considered to be as follows:

$$\alpha_1 = \beta_1 = \alpha_2 = \beta_2 = 5, m_1 = 9, n_1 = 5, p_1 = 7, q_1 = 9, m_2 = 11, n_2 = 9, p_2 = 5, q_2 = 7$$

In this section, uncertainty and external disturbances are added to the system by using multiple harmonics.

$$\begin{aligned} d(t) &= \sin(15t) + 0.7\sin[19(t+4)] + 0.4\sin[28(t+17)] \\ &\quad + 0.08\sin(500t) + 0.03\sin(800t)[\text{Nm}] \end{aligned} \tag{11.30}$$

The block diagram of controlling structure is illustrated in Figure 11.6.

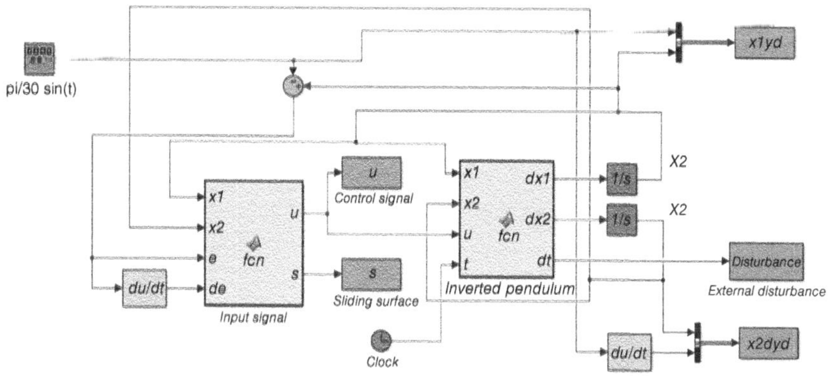

Figure 11.6 Block diagram of fixed-time terminal sliding mode control of inverted pendulum.

The inverted pendulum function is as follows:

```
function [dx1,dx2,dt] = fcn(x1,x2,u,t)

m = 0.1;M = 1;L = 0.5;g = 9.8;

zeros(2,1);

dx1 = x2;

f =((M+m)*g*sin(x1)-m*L*(x2^2)*sin(x1)*cos(x1))/((4/3)*
(M+m)*L-m*L*...
     (0.5+0.5*cos(2*x1)));

g = cos(x1)/((4/3)*(M+m)*L-m*L*(0.5+0.5*cos(2*x1)));

dt = sin(15*t)+0.7*sin(19*(t+4))+0.4*sin(28*(t+17))
+0.08*sin(500*t)+0.03*sin(800*t);

dx2 = f+g*u+dt;
```

The input signal MATLAB function program is as follows:

```
function [u,s] = fcn(x1,x2,e,de)
m = 0.1;M = 1;L = 0.5;g = 9.8;eta = 15;D = 2.21;
m1 = 9;n1 = 5;p1 = 7;q1 = 9;m2 = 11;n2 = 9;p2 = 5;q2 = 7;
alpha1 = 5;alpha2 = 5;beta1 = 5;beta2 = 5;
```

```
T1 = abs(e)^(m1/n1)*sign(e);
T2 = abs(e)^(p1/q1)*sign(e);

T3 = abs(x1)^(m1/n1-1)*sign(x1);
T4 = abs(x1)^(p1/q1-1)*sign(x1);

s = de+alpha1*T1+beta1*T2;

T5 = abs(s)^(m2/n2)*sign(s);
T6 = abs(s)^(p2/q2)*sign(s);

f = ((M+m)*g*sin(x1)-m*L*(x2^2)*sin(x1)*cos(x1))/((4/3)*
(M+m)*L...
    -m*L*(0.5+0.5*cos(2*x1)));

g = cos(x1)/((4/3)*(M+m)*L-m*L*(0.5+0.5*cos(2*x1)));

u = (-1/g)*(f+alpha1*m1/n1*x2*T3+beta1*p1/q1*x2*T4
+alpha2*T5+beta2*T6+...
    (eta+D)*sign(s));
```

Figure 11.7 shows the external disturbance (11.30).

The generated control signal has chattering and is shown in Figure 11.8.

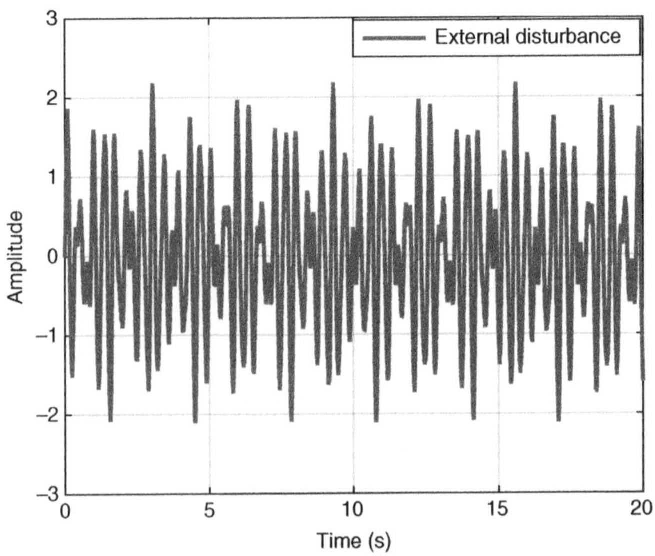

Figure 11.7 External disturbance applied to system dynamics.

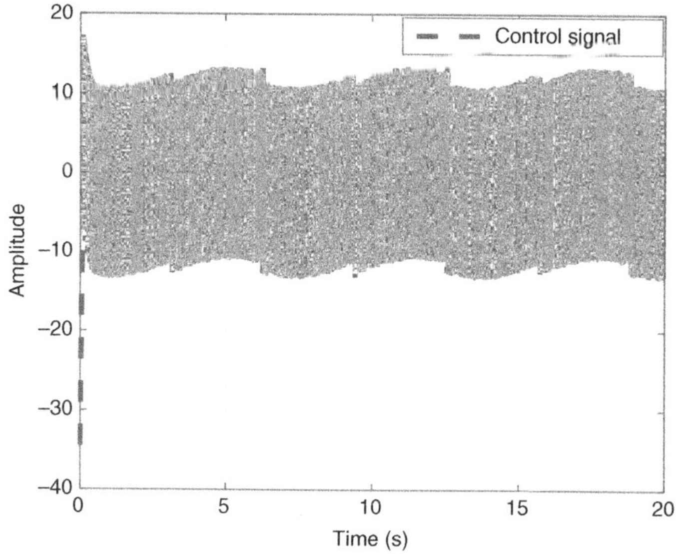

Figure 11.8 Input control signal.

As shown in Figure 11.9, the state variable x_1 follows the desired input signal with acceptable accuracy and speed, and the convergence time in the fixed-time terminal sliding mode control method is improved from 0.53 to 0.43 seconds, comparing to the similar method presented in Figure 11.2.

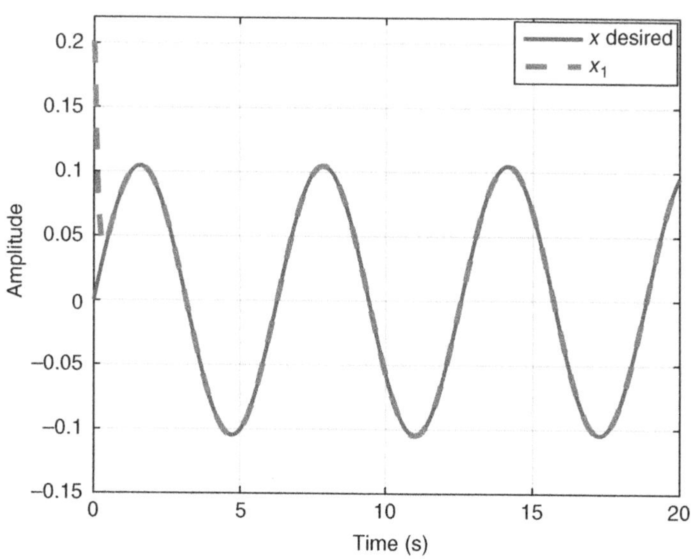

Figure 11.9 Inverted pendulum position tracking using fixed-time TSMC.

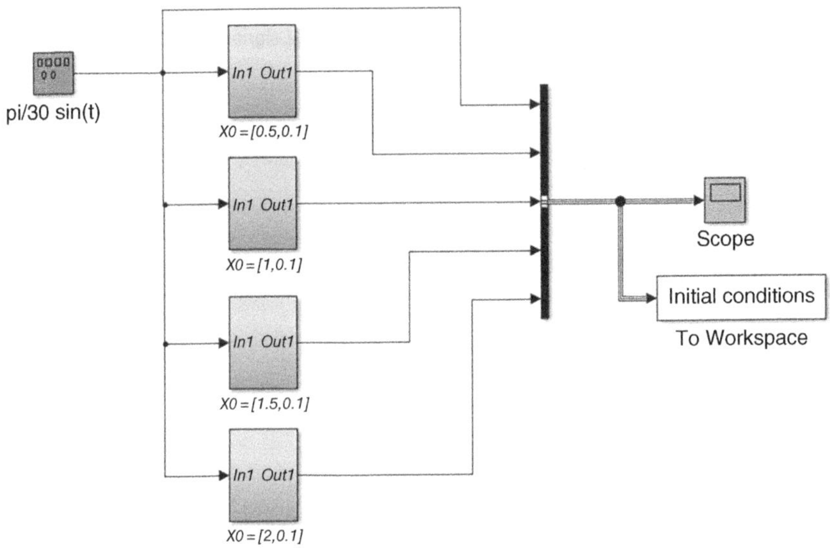

Figure 11.10 Block diagram of the system in Simulink for different initial conditions.

In order to clarify the independence of the initial conditions using fixed-time terminal sliding mode controller, system (11.1) is implemented for different initial conditions as shown in Figure 11.10. According to Figure 11.11, for different initial conditions, the convergence time is constant.

11.3.5 Nonsingular Fixed-Time Terminal Sliding Mode Controller

To avoid singularity phenomena, a nonsingular fixed-time terminal sliding mode controller is proposed in this section. For clear investigation, the singularity phenomenon is examined in the reaching phase and sliding phase.

- **Singularity problem in the reaching phase**

  Consider the control law proved in Section 11.3.3:

$$\lim_{x_1 \to 0} u = -g^{-1}$$
$$\left(f - \ddot{x}_d + \alpha_1 \frac{m_1}{n_1} \tilde{x}_1^{\frac{m_1}{n_1}-1} \tilde{x}_2 + \beta_1 \frac{p_1}{q_1} \tilde{x}_1^{\frac{p_1}{q_1}-1} \tilde{x}_2 + \alpha_2 s^{\frac{m_2}{n_2}} + \beta_2 s^{\frac{p_2}{q_2}} + (\mu + D)\mathrm{sign}(s) \right) = \infty \tag{11.31}$$

where $x \in R$ denotes system states and p, q are odd positive real numbers. Considering that $p_1 < q_1$, in condition of $x_1 = \theta = 0$, an infinite controlling effort will be needed, and it leads to singularity problem. It can also be clarified as follows:

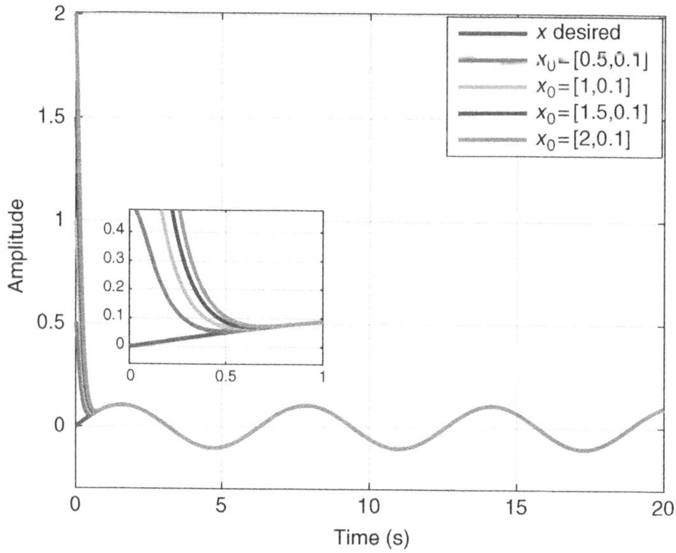

Figure 11.11 Inverted pendulum position tracking using fixed-time TSMC for different initial conditions.

$$\frac{p_1}{q_1} < 1 \Rightarrow \frac{p_1}{q_1} - 1 < 0, \quad \text{so if } x_1 = 0, \quad x_2 \neq 0 \Rightarrow u \to \infty$$

- **Singularity problem in the sliding phase**

  Consider the sliding surface as follows:

$$s = \tilde{x}_2 + \alpha_1 \tilde{x}_1^{\frac{m_1}{n_1}} + \beta_1 \tilde{x}_1^{\frac{p_1}{q_1}}, (\alpha_1, \beta_1) > 0 \tag{11.32}$$

$$s = 0 \Rightarrow \tilde{x}_2 = -\alpha_1 \tilde{x}_1^{\frac{m_1}{n_1}} - \beta_1 \tilde{x}_1^{\frac{p_1}{q_1}} \tag{11.33}$$

The term $\tilde{x}_1^{\frac{p_1}{q_1}-1}$, which caused the singularity phenomenon in the reaching phase, is multiplied by both sides of the above equation.

$$\tilde{x}_1^{\frac{p_1}{q_1}-1} \tilde{x}_2 = -\alpha_1 \tilde{x}_1^{\frac{m_1}{n_1}+\frac{p_1}{q_1}-1} - \beta_1 \tilde{x}_1^{\frac{2p_1}{q_1}-1}$$

Here

$$q_1 < 2p_1 \Rightarrow 1 < \frac{2p_1}{q_1} \Rightarrow 0 < \frac{2p_1}{q_1} - 1$$

According to the above theorem, the term $\tilde{x}_1^{\frac{p_1}{q_1}-1} \tilde{x}_2$ will not create a singularity phenomenon in the sliding phase.

11.3.5.1 Design of Nonsingular Fixed-Time Terminal Sliding Mode Controller

To design the nonsingular fixed-time terminal sliding mode controller the following steps are considered:

- The sliding surface is defined as follows [11]:

$$s = \tilde{x}_1 + \left(\frac{1}{\alpha_1 \tilde{x}_1^{\frac{m_1}{n_1} - \frac{p_1}{q_1}} + \beta_1} \tilde{x}_2 \right)^{\frac{q_1}{p_1}} = \tilde{x}_1 + (\rho \tilde{x}_2)^{\frac{q_1}{p_1}} \tag{11.34}$$

$$\rho(\tilde{x}_1) \triangleq \frac{1}{\alpha_1 \tilde{x}_1^{\frac{m_1}{n_1} - \frac{p_1}{q_1}} + \beta_1} > 0 \tag{11.35}$$

$$\alpha_1 \tilde{x}_1^{\frac{m_1}{n_1} - \frac{p_1}{q_1}} + \beta_1 > 0 \text{ for all } x_1 \in \mathbb{R}.$$

In this section, the stability of the fixed time for the sliding phase is investigated. In the sliding phase the following relation is considered:

$$s = \tilde{x}_1 + \left(\frac{1}{\alpha_1 \tilde{x}_1^{\frac{m_1}{n_1} - \frac{p_1}{q_1}} + \beta_1} \tilde{x}_2 \right)^{\frac{q_1}{p_1}} = 0 \Rightarrow \tilde{x}_1 = - \frac{1}{\left(\alpha_1 \tilde{x}_1^{\frac{m_1}{n_1} - \frac{p_1}{q_1}} + \beta_1 \right)^{\frac{q_1}{p_1}}} \tilde{x}_2^{\frac{q_1}{p_1}} \tag{11.36}$$

$$\begin{aligned} \tilde{x}_2 = -\left(\alpha_1 \tilde{x}_1^{\frac{m_1}{n_1} - \frac{p_1}{q_1}} + \beta_1 \right) \tilde{x}_1^{\frac{p_1}{q_1}} &\Rightarrow \tilde{x}_2 = -\alpha_1 \tilde{x}_1^{\frac{m_1}{n_1}} - \beta_1 \tilde{x}_1^{\frac{p_1}{q_1}} \\ &\Rightarrow \dot{\tilde{x}}_1 = -\alpha_1 \tilde{x}_1^{\frac{m_1}{n_1}} - \beta_1 \tilde{x}_1^{\frac{p_1}{q_1}} \end{aligned} \tag{11.37}$$

According to the fixed-time stability theorem of Lemma 11.2, the above system is fixed-time stable. In other words, the state variables converge to zero at a fixed time and $(x_1, x_2) = (0, 0)$ for all $t > T_1$ and T_1 it is given in the form of the following relationship.

$$T_1 \leq T \frac{1}{\alpha_1} \frac{n_1}{m_1 - n_1} \frac{1}{\beta_1} \frac{q_1}{q_1 - p_{1max}} \tag{11.38}$$

In this section, the stability of the fixed time for the reaching phase is investigated. In the reaching phase, the following relation is considered:

$$s = \tilde{x}_1 + (\rho \tilde{x}_2)^{\frac{q_1}{p_1}} \Rightarrow \dot{s} = \dot{\tilde{x}}_1 + \frac{q_1}{p_1} (\rho \tilde{x}_2)^{\frac{q_1}{p_1} - 1} \left(\dot{\rho} \tilde{x}_2 + \rho \dot{\tilde{x}}_2 \right) \tag{11.39}$$

$$\begin{aligned} \rho(\tilde{x}_1) \triangleq \frac{1}{\alpha_1 \tilde{x}_1^{\frac{m_1}{n_1} - \frac{p_1}{q_1}} + \beta_1} \Rightarrow \dot{\rho}(\tilde{x}_1) &= \frac{-\alpha_1 \left(\frac{m_1}{n_1} - \frac{p_1}{q_1} \right) \tilde{x}_1^{\frac{m_1}{n_1} - \frac{p_1}{q_1} - 1} \dot{\tilde{x}}_1}{\left(\alpha_1 \tilde{x}_1^{\frac{m_1}{n_1} - \frac{p_1}{q_1}} + \beta_1 \right)^2} \\ &= -\alpha_1 \left(\frac{m_1}{n_1} - \frac{p_1}{q_1} \right) \tilde{x}_1^{\frac{m_1}{n_1} - \frac{p_1}{q_1} - 1} \tilde{x}_2 \rho^2 \end{aligned} \tag{11.40}$$

The dynamics of the sliding surface is obtained as follows:

$$\dot{s} = \tilde{x}_2 + \frac{q_1}{p_1}(\rho\tilde{x}_2)^{\frac{q_1}{p_1}-1}\left(-\alpha_1\left(\frac{m_1}{n_1}-\frac{p_1}{q_1}\right)\right)\tilde{x}_1^{\frac{m_1}{n_1}-\frac{p_1}{q_1}-1}\tilde{x}_2{}^2\rho^2 + \rho(f + gu + d - \ddot{x}_d) \tag{11.41}$$

The final goal is to obtain a control law that provides fixed-time stability for the sliding-surface dynamics.

$$\begin{aligned} u = & -\frac{[f + D\,\mathrm{sign}(s)]}{g} + \frac{1}{g\rho}\left[\alpha_1\left(\frac{m_1}{n_1}-\frac{p_1}{q_1}\right)\tilde{x}_1^{\frac{m_1}{n_1}-\frac{p_1}{q_1}-1}\tilde{x}_2{}^2\rho^2 - \frac{p_1}{q_1}\rho^{1-\frac{q_1}{p_1}}\tilde{x}_2{}^{2-\frac{q_1}{p_1}}\right]\ldots \\ & -\frac{1}{g\rho}\frac{p_1}{q_1}\rho^{1-\frac{q_1}{p_1}}\tilde{x}_2{}^{1-\frac{q_1}{p_1}}\left(\alpha_2 s^{\frac{m_2}{n_2}} + \beta_2 s^{\frac{p_2}{q_2}}\right) \end{aligned} \tag{11.42}$$

The control parameters are as follows:

$$\alpha_1, \beta_1 > 0, m_1 > n_1, p_1 < q_1 < 2p_1, \frac{m_1}{n_1} - \frac{p_1}{q_1} > 1, \quad \alpha_2, \beta_2 > 0, m_2 > n_2, p_2 < q_2 \tag{11.43}$$

According to the values of the above parameters, the term $x_2{}^{1-\frac{q_1}{p_1}}$ has a negative power and while it is $x_2 = 0$, it can cause a singularity problem. To avoid this problem, a coefficient μ_τ is added in the control law as follows:

$$\begin{aligned} u = & -\frac{[f + D\,\mathrm{sign}(s)]}{g} + \frac{1}{g\rho}\left[\alpha_1\left(\frac{m_1}{n_1}-\frac{p_1}{q_1}\right)\tilde{x}_1^{\frac{m_1}{n_1}-\frac{p_1}{q_1}-1}\tilde{x}_2{}^2\rho^2 - \frac{p_1}{q_1}\rho^{1-\frac{q_1}{p_1}}\tilde{x}_2{}^{2-\frac{q_1}{p_1}}\right]\ldots \\ & -\frac{1}{g}\frac{p_1}{q_1}\rho^{-\frac{q_1}{p_1}}\mu_\tau\tilde{x}_2{}^{1-\frac{q_1}{p_1}}\left(\alpha_2 s^{\frac{m_2}{n_2}} + \beta_2 s^{\frac{p_2}{q_2}}\right) \end{aligned} \tag{11.44}$$

$$\mu_\tau\left(\tilde{x}_2^{\frac{q_1}{p_1}-1}\right) = \begin{cases} \sin\left(\dfrac{\pi}{2}\dfrac{\tilde{x}_2^{\frac{q_1}{p_1}-1}}{\tau}\right) & \text{if } \tilde{x}_2^{\frac{q_1}{p_1}-1} \le \tau \\ 1 & \text{Otherwise} \end{cases}, \mu_\tau > 0 \quad \text{if} \quad \tilde{x}_2 \neq 0 \tag{11.45}$$

τ is a positive constant.

Placing u in the sliding-surface dynamics, the following equation yields:

$$\dot{s} = -\mu_\tau\left(\alpha_2 s^{\frac{m_2}{n_2}} + \beta_2 s^{\frac{p_2}{q_2}}\right) - \frac{q_1}{p_1}\rho^{\frac{q_1}{p_1}}\tilde{x}_2^{\frac{q_1}{p_1}-1}[d + D\,\mathrm{sign}(s)] \tag{11.46}$$

Theorem 11.4
Controller (11.44) is a fixed-time stable controller for system (11.1).

Proof: To prove the fixed-time stability, the following Lyapunov function is considered

$$V(s) = \frac{1}{2}s^2 \Rightarrow \dot{V}(s) = s\dot{s} = s\left(-\mu_\tau\left(\alpha_2 s^{\frac{m_2}{n_2}} + \beta_2 s^{\frac{p_2}{q_2}}\right) - \frac{q_1}{p_1}\rho^{\frac{q_1}{p_1}}\tilde{x}_2^{\frac{q_1}{p_1}-1}[d + D\,\mathrm{sign}(s)]\right) \tag{11.47}$$

$$\begin{aligned}\dot{V} &\le -\mu_\tau\alpha_2 s^{\frac{m_2+n_2}{2n_2}} - \mu_\tau\beta_2 s^{\frac{p_2+q_2}{q_2}} + \frac{q_1}{p_1}\rho^{\frac{q_1}{p_1}}\tilde{x}_2^{\frac{q_1}{p_1}-1}\\ (|d| - D)|s| &\le -\mu_\tau\alpha_2\left(s^2\right)^{\frac{m_2+n_2}{2n_2}} - \mu_\tau\beta_2\left(s^2\right)^{\frac{p_2+q_2}{2q_2}} = -\mu_\tau\alpha_2(2V)^{\frac{m_2+n_2}{2n_2}} - \mu_\tau\beta_2(2V)^{\frac{p_2+q_2}{2q_2}}\end{aligned} \tag{11.48}$$

$$y = \sqrt{2V} \Rightarrow \dot{y} = \frac{2\dot{V}}{2\sqrt{2V}} = \frac{\dot{V}}{\sqrt{2V}} \Rightarrow \dot{V} = \sqrt{2V}\dot{y}$$

$$\begin{aligned}&\sqrt{2V}\dot{y} \le -\mu_\tau\alpha_2\left(\sqrt{2V}\right)^{\frac{m_2+n_2}{n_2}} - \mu_\tau\beta_2\left(\sqrt{2V}\right)^{\frac{p_2+q_2}{q_2}}\\ &y\dot{y} \le -\mu_\tau\alpha_2(y)^{\frac{m_2+n_2}{n_2}} - \mu_\tau\beta_2(y)^{\frac{p_2+q_2}{q_2}} \Rightarrow \dot{y} \le -\mu_\tau\alpha_2(y)^{\frac{m_2}{n_2}} - \mu_\tau\beta_2(y)^{\frac{p_2}{q_2}}\end{aligned} \tag{11.49}$$

According to the fixed-time stability theorem, the above system is fixed time stable and the reaching phase is realized in a fixed time.

The phase plane is divided into two parts S_1, S_2 as shown in Figure 11.12 [11, 14].

In the region S_1, the following condition is satisfied, and it represents the regions where singularity does not occur.

$$S_1 = \left\{(x_1, x_2)\middle| x_2^{\frac{q_1}{p_1}-1} \ge \tau\right\} \tag{11.50}$$

In the region S_2, the following condition is satisfied, and it represents the regions where singularity may happen.

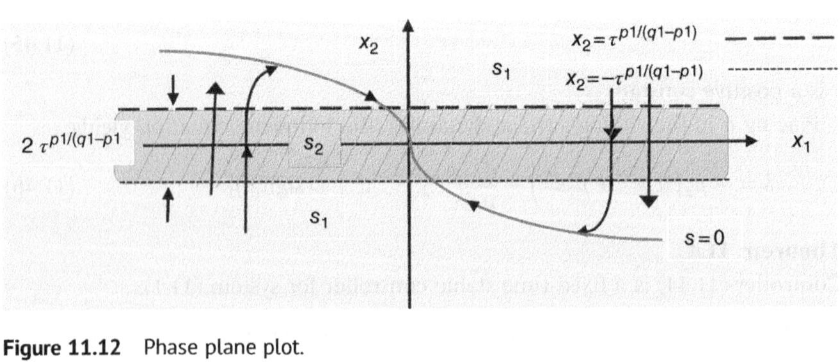

Figure 11.12 Phase plane plot.

$$S_2 = \left\{(x_1, x_2)\middle| x_2^{\frac{q_1}{p_1}-1} < \tau\right\} \tag{11.51}$$

Here:

$$x_2^{\frac{q_1}{p_1}-1} = x_2^{\frac{q_1-p_1}{p_1}} = \tau \Rightarrow x_2 = \tau^{\frac{p_1}{q_1-p_1}} \tag{11.52}$$

In the following, the stability is investigated in each region S_1, S_2.

- Region S_1

  Consider the following equation:

$$\dot{y} \leq -\mu_\tau \alpha_2 (y)^{\frac{m_2}{n_2}} - \mu_\tau \beta_2 (y)^{\frac{p_2}{q_2}} \tag{11.53}$$

  Region S_1 was as follows:

$$S_1 = \left\{(x_1, x_2)\middle| x_2^{\frac{q_1}{p_1}-1} \geq \tau\right\} \tag{11.54}$$

where

$$\mu_\tau\left(x_2^{\frac{q_1}{p_1}-1}\right) = \begin{cases} \sin\left(\dfrac{\pi}{2}\dfrac{x_2^{\frac{q_1}{p_1}-1}}{\tau}\right) & \text{if} \quad x_2^{\frac{q_1}{p_1}-1} \leq \tau \\ 1 & \text{Otherwise} \end{cases} \tag{11.55}$$

  Finally, $\dot{y}$ is rewritten as follows:

$$\dot{y} \leq -\alpha_2 (y)^{\frac{m_2}{n_2}} - \beta_2 (y)^{\frac{p_2}{q_2}} \tag{11.56}$$

  The modes (x_1, x_2) in region S_1 reach the sliding surface $s = 0$ in fixed time $t < T_2$. The upper bound of the convergence time is calculated from the following equation:

$$T_2 = \frac{1}{\alpha_2}\frac{n_2}{m_2 - n_2} + \frac{1}{\beta_2}\frac{q_2}{q_2 - p_2} \tag{11.57}$$

- Region S_2

$$S_2 = \left\{(x_1, x_2)\middle| x_2^{\frac{q_1}{p_1}-1} < \tau\right\} \Rightarrow 0 < \mu_\tau < 1 \quad \text{when} \quad x_2 \neq 0 \tag{11.58}$$

  If the time to pass through the area S_2 is denoted by $\varepsilon(\tau)$, which depends on the boundary width $2\tau^{\frac{p_1}{q_1-p_1}}$, the modes (x_1, x_2) go through the reaching phase at a fixed time $t < T_2 + \varepsilon(\tau)$ and reach the sliding surface $s = 0$.

To prove non-absorbing axis $x_1(x_2 = 0)$ except at point$(x_1, x_2) = (0, 0)$, consider the control signal as follows:

$$u = -\frac{[f + D\,\mathrm{sign}(s)]}{g} + \frac{1}{g\rho}\left[\alpha_1\left(\frac{m_1}{n_1} - \frac{p_1}{q_1}\right)\tilde{x}_1^{\frac{m_1}{n_1}-\frac{p_1}{q_1}-1}\tilde{x}_2^{2}\rho^{2} - \frac{p_1}{q_1}\rho^{1-\frac{q_1}{p_1}}\tilde{x}_2^{2-\frac{q_1}{p_1}}\right]\ldots$$
$$-\frac{1}{g}\frac{p_1}{q_1}\rho^{-\frac{q_1}{p_1}}\mu_\tau\tilde{x}_2^{1-\frac{q_1}{p_1}}\left(\alpha_2 s^{\frac{m_2}{n_2}} + \beta_2 s^{\frac{p_2}{q_2}}\right) \tag{11.59}$$

$$\lim_{x_2\to 0}\mu_\tau x_2^{1-\frac{q_1}{p_1}} = \lim_{x_2\to 0}\sin\left(\frac{\pi}{2\tau}x_2^{\frac{q_1}{p_1}-1}\right)x_2^{1-\frac{q_1}{p_1}} = \lim_{x_2\to 0}\frac{\sin\left(\frac{\pi}{2\tau}x_2^{\frac{q_1}{p_1}-1}\right)}{x_2^{\frac{q_1}{p_1}-1}} = \frac{\pi}{2\tau} \tag{11.60}$$

Therefore, when $x_2 = 0$, the control signal is rewritten as follows:

$$u = -\frac{[f + \gamma\,\mathrm{sign}(s)]}{g} - \frac{1}{g}\frac{p_1}{q_1}\frac{\pi}{2\tau}\rho^{-\frac{q_1}{p_1}}\left(\alpha_2 s^{\frac{m_2}{n_2}} + \beta_2 s^{\frac{p_2}{q_2}}\right) \tag{11.61}$$

Substituting control signal yields:

$$\dot{x}_2 = d - D\,\mathrm{sign}(s) - \frac{p_1}{q_1}\frac{\pi}{2\tau}\rho^{-\frac{q_1}{p_1}}\left(\alpha_2 s^{\frac{m_2}{n_2}} + \beta_2 s^{\frac{p_2}{q_2}}\right), D \geq |d|, \rho > 0, \tau > 0 \tag{11.62}$$

The following two cases are considered.

$$\begin{aligned} s > 0 &\Rightarrow \dot{x}_2 < 0 \\ s < 0 &\Rightarrow \dot{x}_2 > 0 \end{aligned} \tag{11.63}$$

The above relationships show that $x(t)$ enters region $\boldsymbol{S_1}$ from region $\boldsymbol{S_2}$ in a fixed time and no singularity occurs in this region.

$$u = -\frac{[f + D\ \mathrm{sign}(s)]}{g} + \frac{1}{g\rho}\left[\alpha_1\left(\frac{m_1}{n_1} - \frac{p_1}{q_1}\right)\tilde{x}_1^{\frac{m_1}{n_1}-\frac{p_1}{q_1}-1}\tilde{x}_2^{2}\rho^{2} - \frac{p_1}{q_1}\rho^{1-\frac{q_1}{p_1}}\tilde{x}_2^{2-\frac{q_1}{p_1}}\right]\ldots$$
$$-\frac{1}{g}\frac{p_1}{q_1}\rho^{-\frac{q_1}{p_1}}\left(\alpha_2 s^{\frac{m_2}{n_2}} + \beta_2 s^{\frac{p_2}{q_2}}\right), x_2(0) = 0$$
$$u = -\frac{[f + D\ \mathrm{sign}(s)]}{g} + \frac{1}{g\rho}\left[\alpha_1\left(\frac{m_1}{n_1} - \frac{p_1}{q_1}\right)\tilde{x}_1^{\frac{m_1}{n_1}-\frac{p_1}{q_1}-1}\tilde{x}_2^{2}\rho^{2} - \frac{p_1}{q_1}\rho^{1-\frac{q_1}{p_1}}\tilde{x}_2^{2-\frac{q_1}{p_1}}\right]\ldots$$
$$-\frac{1}{g}\frac{p_1}{q_1}\rho^{-\frac{q_1}{p_1}}\mu_\tau\tilde{x}_2^{1-\frac{q_1}{p_1}}\left(\alpha_2 s^{\frac{m_2}{n_2}} + \beta_2 s^{\frac{p_2}{q_2}}\right), x_2(0) \neq 0 \tag{11.64}$$

The control parameters for the simulation are presented as follows. These parameters satisfy the conditions (11.43).

$$\alpha_1 = \alpha_2 = \beta_1 = \beta_2 = 5$$
$$m_1 = 9, n_1 = 5, m_2 = 11, n_2 = 9, p_1 = 7, q_1 = 9, p_2 = 5, q_2 = 7, \tau = 0.1 \tag{11.65}$$

The upper bound of the sliding phase convergence time is as follows:

$$T_1 = \frac{1}{\alpha_1}\frac{n_1}{m_1 - n_1} + \frac{1}{\beta_1}\frac{q_1}{q_1 - p_1} = 2.8750 \tag{11.66}$$

The upper bound of the reaching phase convergence time is as follows:

$$T_2 = \frac{1}{\alpha_2}\frac{n_2}{m_2 - n_2} + \frac{1}{\beta_2}\frac{q_2}{q_2 - p_2} = 4 \tag{11.67}$$

11.3.5.2 MATLAB Programming

In this section, the simulation of the inverted pendulum system with the fixed time nonsingular terminal sliding mode method is investigated. The values of the parameters are similar to the previous programs. Also, the codes related to the functions are presented below. The program has two controllers introduced in (11.64). These controllers are designed for some initial conditions of the x_2 state variable that would cause the singularity problem.

```
close all; clear all; clc;
warning off
%% Initialization
Tf = 20; Ts = 0.001; time = 0:Ts:Tf; Nt = numel(time);
x = zeros(2,Nt); u = zeros(1,Nt); x(:,1) = [1 0];
m = 0.1; M = 1; L = 0.5; g = 9.8;
%% Design parameters
m1 = 9; n1 = 5; p1 = 7; q1 = 9; m2 = 11; n2 = 9; p2 = 5;
q2 = 7; alpha1 = 5; alpha2 = 5; beta1 = 5; beta2 = 5; D =
0.1; tau = 0.1;
%% Design process
for k = 1:Nt-1
f(k) = ((M+m)*g*sin(x(1,k))-
m*L*(x(2,k)^2)*sin(x(1,k))*cos(x(1,k)))/((4/3)...
     *(M+m)*L-m*L*(0.5+0.5*cos(2*x(1,k))));
b(k) = cos(x(1,k))/((4/3)*(M+m)*L-m*L*(0.5+0.5*cos(2*x(1,
k))));
d(k) = 0.1*sin(time(k));
ro(k) = 1/(alpha1*x(1,k)^(m1/n1-p1/q1)+beta1);
```

```
s(k) = x(1,k)+(ro(k)*x(2,k))^(q1/p1);
if x(2,k)^(q1/p1-1)<tau
    mu = sin((pi/2*tau)*(x(2,k)^(q1/p1-1)));
else
    mu = 1;
end
 u(k) = -(f(k)+D*sign(s(k)))/b(k)+(1/b(k)*ro(k))*(alpha1*
(m1/n1-p1/q1)*...
       x(1,k)^(m1/n1-p1/q1-1)*x(2,k)^2*ro(k)^2-(p1/q1)*ro
(k)*(1-q1/p1)*...
       x(2,k)^(2-q1/p1))-(p1/b(k)*q1)*ro(k)^(-q1/p1)*
(alpha2*s(k)^(m2/n2)...
       +beta2*s(k)^(p2/q2));    %for x2(0) = 0

%  u(k) = -(f(k)+D*sign(s(k)))/b(k)+(1/b(k)*ro(k))*
(alpha1*(m1/n1-p1/q1)*...
%       x(1,k)^(m1/n1-p1/q1-1)*x(2,k)^2*ro(k)^2-(p1/q1)*ro
(k)*(1-q1/p1)*...
%       x(2,k)^(2-q1/p1))-(p1/b(k)*q1)*ro(k)^(-q1/p1)*mu*x
(2,k)^(1-q1/p1)*...
%        (alpha2*s(k)^(m2/n2)+beta2*s(k)^(p2/q2)); %for
x2(0)≠ 0
%% model dynamics
 x(1,k+1) = x(1,k)+Ts*x(2,k);
 x(2,k+1) = x(2,k)+Ts*(f(k)+b(k)*u(k)+d(k));
end
%% plot
figure(1)
subplot(2,1,1)
plot(time,x(1,:))
xlabel('Time (s)')
ylabel('x 1^(t)')
grid
subplot(2,1,2)
plot(time,x(2,:))
xlabel('Time(sec)')
ylabel('x_2(t)')
grid
```

As can be seen in Figure 11.13, the state variables x_1 and x_2 converge to zero at fixed time.

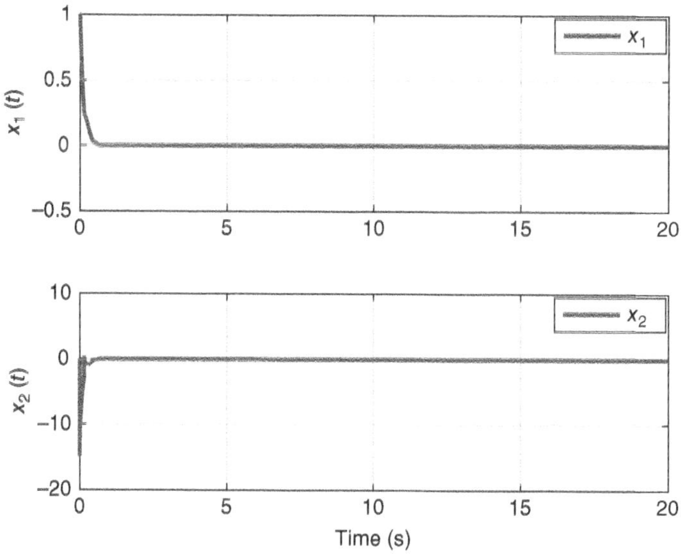

Figure 11.13 Behavior of state variables.

References

1 Youhao, H., Wang, H., He, S. et al. (2021). Adaptive tracking control of an electronic throttle valve based on recursive terminal sliding mode. *IEEE Transactions on Vehicular Technology* 70 (1): 251–262. https://doi.org/10.1109/TVT.2020.3045778.

2 Kunfeng, L. and Xia, Y. (2013). Adaptive attitude tracking control for rigid spacecraft with finite-time convergence. *Automatica* 49: 3591–3599. https://doi.org/10.1016/j.automatica.2013.09.001.

3 Moghaddam, R.K. and Mohamadie, N.K. (2023). Fast terminal nonsingular sliding mode control of wave energy converters for maximum power absorption. *Ocean Engineering* 281. https://doi.org/10.1016/j.oceaneng.2023.114473.

4 Polyakov, A. (2012). Nonlinear feedback design for fixed-time stabilization of linear control systems. *IEEE Transactions on Automatic Control* 57 (8): 2106–2110.

5 Zhong, C., Liping, W., Guo, J. et al. (2018). Robust adaptive attitude manoeuvre control with finite-time convergence for a flexible spacecraft. *Transactions of the Institute of Measurement and Control* 40 (2): 425–435. https://doi.org/10.1177/0142331216659337.

6 Zhang, Y., Sun, M., and Chen, Z. (2012). Finite-time convergent guidance law with impact angle constraint based on sliding-mode control. *Nonlinear Dynamics* 70: 619–625. https://doi.org/10.1007/s11071-012-0482-3.

7 ShuangheYu, X.Y., Zadeh, B.S., and Man, Z. (2005). Continuous finite-time control for robotic manipulators with terminal sliding mode. *Automatica* 41: 1957–1964.

8 Zuo, Z. (2015). Nonsingular fixed-time consensus tracking for second-order multi-agent networks. *Automatica* 54: 305–309.

9 Huang, Y. and Jia, Y. (2018). Robust adaptive fixed-time tracking control of 6-DOF spacecraft fly-around mission for noncooperative target. *International Journal of Robust and Nonlinear Control* 28 (6): 2598–2618.

10 Ni, J. (2022). Fixed-time terminal sliding mode tracking protocol design for high-order multiagent systems with directed communication topology. *ISA Transactions* 124: 444–457.

11 Zuo, Z. (2015). Non-singular fixed-time terminal sliding mode control of non-linear systems. *IET Control Theory and Applications* 9 (4): 545–552.

12 Tian, Y., Cal, Y., and Deng, Y. (2020). A fast nonsingular terminal sliding mode control method for nonlinear systems with fixed-time stability guarantees. *IEEE Access* 8: 60444–60454. https://doi.org/10.1109/ACCESS.2020.2980044.

13 Zhang, J., Shuanghe, Y., Defeng, W., and Yan, Y. (2020). Nonsingular fixed-time terminal sliding mode trajectory tracking control for marine surface vessels with anti-disturbances. *Ocean Engineering* 217. https://doi.org/10.1016/j.oceaneng.2020.108158.

14 Khanzadeh, A. and Mohammadzaman, I. (2023). Continuous fixed-time nonsingular terminal sliding mode control of second-order nonlinear systems with matched and mismatched disturbances. *Scientia Iranica, Transactions D: Computer Science & Engineering and Electrical Engineering* 30: 142–153. https://doi.org/10.24200/sci.2021.56129.4563.

12

Event-Triggering Sliding Mode Controllers

12.1 Introduction

Event-based controllers consist of two parts: a feedback controller that calculates the control signal and a trigger mechanism that indicates when the control signal should be updated. Event-triggering controllers are a type of reactive controller. In such controllers, the condition for triggering is determined based on the continuous measurements of the states. Event-based controllers are used both in mechanisms with continuous and discrete time structures. In this chapter, event-triggering sliding mode controllers are designed and analyzed for linear and nonlinear structures and, finally, the MATLAB programming and its related simulation results are proposed [1].

12.2 Designing Event-Triggering Sliding Mode Controller for Linear Systems

Consider the following LTI system with matched disturbance d:

$$\dot{x} = Ax + B(u + d) \quad ; \quad \|d\| \le d_{max} \tag{12.1}$$

Although event-triggered systems are not discrete time systems, the controlling signal is considered to be fixed in time interval $t \in [t_i \ . \ t_{i+1})$, and the below control is considered:

$$u(t) = -\left(c^T B\right)^{-1}\left(c^T Ax(t_i) + K \operatorname{sign} s(t_i)\right) \tag{12.2}$$

In the above relation, it is assumed that the set of t_i s is the set of discrete times that at each of the starting points ($t_1 \ . \ t_2 \ . \ ....$), the control signal can be applied to the system. The system receives the controller value at each moment of t_i and

Methods of Developing Sliding Mode Controllers: Design and MATLAB Simulation, First Edition.
Reihaneh Kardehi Moghaddam and Mostafa Rabbani.

Published 2025 by John Wiley & Sons, Inc.

remains constant until the next moment, and the time interval between two events is represented by T_i.

In event-based systems, the error is considered as $e(t) = x(t_i) - x(t)$ while in systems with continuous controller, since $x(t_i)$ and $x(t)$ are always equal, the error value is zero. However, the event-based controller is not continuous and must reduce this error to zero [2].

12.2.1 Definitions and Mathematical Preliminaries

Definition 12.1

An event-triggered system with sliding mode controller is globally stable if

$$\inf_{\substack{x \in \mathbb{R}^n \\ i \in \mathbb{Z}_{\geq 0} \\ |d(t)| \leq d_0}} T_i > a_T \tag{12.3}$$

where a_T is a positive constant and $T_i = t_{i+1} - t_i$ is the triggering time interval [3, 4].

Definition 12.2

An event-triggered system with sliding mode controller (12.2) is semi-globally robust stable [5, 6] if for every subset $D \subset \mathbb{R}^n$

$$\inf_{\substack{x \in \mathcal{D} \subset \mathbb{R}^n \\ i \in \mathbb{Z}_{\geq 0} \\ |d(t)| \leq d_0}} T_i > a_T \tag{12.4}$$

Definition 12.3

An event-triggered system with sliding mode controller (12.2) is locally robust stable if for some subset D in $\mathcal{K} \subset \mathbb{R}^n$

$$\inf_{\substack{x \in \mathcal{D} \cap \mathcal{K}\ ,\ \mathcal{D} \subset \mathcal{K} \subset \mathbb{R}^n \\ i \in \mathbb{Z}_{\geq 0} \\ |d(t)| \leq d_0}} T_i > a_T \tag{12.5}$$

Definition 12.4

The system (12.1) is said to be a practical sliding mode controller if for any $\varepsilon > 0$ there exists a $t_1 \in [t_0\ .\ \infty)$ such that the sliding trajectories remain bounded in the neighborhood of sliding manifold with the bound given by ε for all time $t \geq t_1$.

Lemma 12.1

Consider the system (12.1) and the sliding manifold $S = \{x \in R^n \mid s = c^T x = 0\}$.

Suppose the following inequality for a given $\alpha \geq 0$:

$$\|c\|\ \|A\|\ \|e(t)\| < \alpha \text{ for all } t \geq 0 \tag{12.6}$$

If the gain K is chosen such that $K > sup_{t \geq 0}|c^T Bd(t)| + \alpha$, then the practical sliding mode controller is executed as (12.2) [4].

Suppose that the condition (12.6) is valid for all times, accordingly it is concluded that:

$$|s(t_i) - s(t)| = |c^T x(t_i) - c^T x(t)| \leq \|c\| \, \|e(t)\| < \frac{\alpha}{\|A\|} \tag{12.7}$$

The above relation expresses the maximum distance that sliding paths can take from its previous sampled value. The value of α should be chosen to satisfy the constraint. The larger the value of α, the greater the range of stability.

12.2.2 Controller Design

Considering the following global event-triggering scheme:

$$t_{i+1} = \inf\{t > t_i : \|c\| \, \|A\| \, \|e(t)\| \geq \sigma(\|x(t_i)\| + \alpha)\} \quad \alpha \in (0. \infty) \; \sigma \in (0.1) \tag{12.8}$$

It is guaranteed that the system trajectories remain bounded within a sliding bound in the vicinity of sliding manifold $s = c^T x$.

In event-triggered sliding mode controllers, the switching coefficient of the controller is always set based on event-based parameters. In the structure of semi-globally stable event-based controllers, the excitation law is only dependent on a series of fixed parameters; as a result, the switching coefficient in the sliding mode controller is adjusted based on these fixed parameters. In event-based controllers, the switching rules depend on the sampled values of state $x(t_i)$. The event-based sliding control law is presented based on the following relationship, which holds for all time $t \in [t_i. \, t_{i+1})$ as follows [4, 5]:

$$u(t) = -\left(c^T B\right)^{-1}\left(c^T A x(t_i) + K(x(t_i)) \text{sign } s(t_i)\right) \tag{12.9.1}$$

where

$$t_{i+1} = \inf\{t > t_i : \|c\| \, \|A\| \, \|e(t)\| \geq \|x(t_i)\| + \alpha\} \tag{12.9.2}$$

As it is mentioned K is a function of $x(t_i)$ and should be determined in the way that satisfies the following inequalities:

a. $K(x(t_i)) > sup_{t \geq 0}\left|C^T Bd(t)\right|$ (12.10)

b. $K(x(t_i)) > \|x(t_i)\| + \alpha$ for all $x \in \mathbb{R}^n$ (12.11)

One of the simplest structures that can be considered for the switching coefficient is

$$K(x(t_i)) = K_1 + K_2\|x(t_i)\|\|\|, \tag{12.12}$$

in which $K_1 > |c^T B| d_0 + \alpha$ and $K_2 > 0$.

Theorem 12.1

Consider system (12.1). Let $\alpha \in (0.\ \infty)$ satisfy $\|c\|\ \|A\|\ \|e(t)\| < \|x(t_i)\| + \alpha$, then the controlling law (12.9) guarantees the system's trajectories remain within the bound

$$\left\{x \in \mathbb{R}^n : \left|c^T x\right| \leq \frac{\alpha + \|x(t_i)\|}{\|A\|}\right\} \tag{12.13}$$

where K is chosen as follows:

$$K(x(t_i)) > \sup_{t \geq 0}\left|c^T Bd(t)\right| + \|x(t_i)\| + \alpha \tag{12.14}$$

Proof: Considering Lyapunov function $V = \frac{1}{2}s^2$ yields:

$$\begin{aligned}
\dot{V}(s(t)) &= s(t)\dot{s}(t) \\
&= s(t)\left(c^T Ax(t) + c^T Bu(t_i) + c^T Bd(t)\right) \\
&= s(t)\left(-c^T Ae(t) - K_1 \operatorname{sign} s(t_i) - K_2\|x(t_i)\|\operatorname{sign} s(t_i) + c^T Bd(t)\right)
\end{aligned}$$

substituting $e(t) = x(t_i) - x(t)$ in the above equation yields:

$$\begin{aligned}
\dot{V}(s(t)) &= -s(t)c^T Ae(t) - s(t)K_1 \operatorname{sign} s(t_i) - s(t)K_2\|x(t_i)\|\operatorname{sign} s(t_i) + s(t)c^T Bd(t) \\
&\leq |s(t)|\left|c^T Ae(t)\right| - s(t)K_1 \operatorname{sign} s(t_i) - s(t)K_2\|x(t_i)\|\operatorname{sign} s(t_i) + |s(t)|\left|c^T Bd(t)\right| \\
&< |s(t)|\ \|x(t_i)\| + |s(t)|\alpha - s(t)K_1 \operatorname{sign} s(t_i) - s(t)K_2\|x(t_i)\|\operatorname{sign} s(t_i) + |s(t)|\left|c^T B\right|d_{max}
\end{aligned}$$

If the system trajectories start from a region with sign $s(t_i) =$ sign $s(t)$, the above relation will be rewritten as follows.

$$\begin{aligned}
\dot{V}(s(t)) &< -|s(t)|K_1 - |s(t)|K_2\|x(t_i)\| + |s(t)|\ \|x(t_i)\| + |s(t)|\alpha + |s(t)|\left|c^T B\right|d_{max} \\
&= -|s(t)|\left(K_1 - \alpha + \left|c^T B\right|d_{max}\right) - |s(t)|(K_2 - 1)\ \|x(t_i)\| \\
&\leq -|s(t)|\left(K_1 - \alpha + \left|c^T B\right|d_{max}\right) \leq -\eta|s|
\end{aligned}$$

where $\eta > 0$ and $K_1 > |c^T B|d_0 + \alpha + \eta$. It is concluded that the system trajectories are attracted toward the sliding surface as long as sign $s(t_i) =$ sign $s(t)$. This holds for all triggering instants. As a result, the trajectories reach the sliding surface due to $\dot{V} < 0$ and enter the region where the sign of $s(t)$ changes before next triggering occurs. So, in time interval the decrease of V cannot be guaranteed due to zero crossing of s. However, the sliding trajectories are ultimately bounded within this region since $\dot{V}(s(t)) < 0$ outside the region. Therefore, the ultimate bound is the region where sign $s(t_i) =$ sign $s(t)$. The maximum deviation of sliding variable from its immediate sampled is:

$$|s(t_i) - s(t)| = \left|c^T x(t_i) - c^T x(t)\right| \leq \|c\|\ \|e(t)\| \leq \frac{\alpha + \|x(t_i)\|}{\|A\|}$$

Then, the maximum value of bound can be obtained for the case $s(t_i) = 0$ from (12.13) and the proof is completed. □

The ultimately boundedness of trajectories is also proved in [4].

12.3 MATLAB Programming and Simulation Results: Event-Triggering Sliding Mode Controller for Linear Systems

The following linear system is considered [5]:

$$\dot{x} = \begin{bmatrix} 0 & 1 \\ 4 & 5 \end{bmatrix} x + \begin{bmatrix} 0 \\ 1 \end{bmatrix} u, \quad d = 0.1\sin(10t), \quad d_{max} = 0.1 \tag{12.15}$$

The controlling parameters are defined as:

$$s = c^T x = [0.5\ 1]x, \quad K_1 = 1.2, K_2 = 1.5, \sigma = 0.85, \alpha = 0.5$$

the initial value of states is considered to be

$$x_0 = [0.5,1]$$

The event-triggered SMC is calculated from (12.9.1) and (12.9.2), and the following MATLAB programming is considered.

```
%initialization
x0 = [2; 1];
tf = 10;deltat = 0.0001;
%system parameters
c = [0.5;1];
k1 = 1.2; k2 = 1.1;
a = 0.5; sig = 0.85;
%LTI system
A = [0 1;4 5]; B = [0;1];
%initialize arrays
xindex = [];uindex = [];tiindex = [];IVindex = [];
%intereventindex
x = x0;xi = x0;counter = 0;
xindex = [xindex x0];
%sliding surface and controller initial values
    s = c'*x;
    k = k1 + k2 * norm(x);
    u = -inv(c'*B)*(c'*A*x+k*sign(s));
    uindex = [uindex u];intereventindex = [];
```

```
%extracting event-triggered SMC
for i = 1: (tf/deltat)-1
    x = deltat.*( A*x + B*(u+0.1*sin(10*i*deltat)))+x;
%eq (12.15)
    e = xi-x;
    lhs = norm(c)*norm(A)*norm(e);
    rhs = sig * (norm(xi)+a);

    if lhs >= rhs %eq(12.9.2)
    xi = x;
    s = c'*xi;
    k = k1 + k2 * norm(xi);
    ui = -inv(c'*B)*(c'*A*xi+k*sign(s));%eq(12.9.1)
    u = ui;
    tiindex = [tiindex i*deltat];% the triggering time
    end
   uindex = [uindex u];
   xindex = [xindex x];
end
IVindex = [tiindex 0]-[0 tiindex]; %the interevent time is
ti+1-ti




t = 0:deltat:tf-deltat;





%plot the results
t = 0:deltat:tf-deltat;
figure(1)
plot(t,xindex(1,:),t,xindex(2,:),'linewidth',2)
xlabel('t(s)');
ylabel('x(t)');
legend('x1','x2')
figure(2)
plot(xindex(1,:),xindex(2,:),'linewidth',2)
```

```
title('phase portraite');
xlabel('x1');
ylabel('x2');
figure(3)
plot(t,uindex(1,:),'linewidth',2)
xlabel('t(s)');
ylabel('u(t)');
figure(4)
plot(tiindex(1,:),IVindex(1,1:end-1),'*')
xlabel('t(s)');
ylabel('triggering time interval(t)');
```

The simulation results shown in Figure 12.1 indicate the efficiency of controller in keeping the states bounded in a neighborhood of sliding surface.

In addition Figure 12.2a illustrates the event-triggered controlling signal, and Figure 12.2b shows the intervals as it is defined in definition 12.11.

12.4 Design of Event-Triggering SMC for Nonlinear Systems

Consider the following nonlinear system:

$$\begin{aligned}\dot{x}_1 &= f_1(x_1) + B_1 x_2 \\ \dot{x}_2 &= f_2(x_1.x_2) + B_2 u + B_2 d\end{aligned} \tag{12.16}$$

where $f_1(.)$ and $f_2(.)$ are two nonlinear functions satisfying Lipshitz conditions and $x_1 \in \mathbb{R}^{n-1}$ and $x_2 \in \mathbb{R}$. The disturbance d is bounded as $|d| \leq d_0$. It is considered that $f_1(.)$ has a unique equilibrium point and $f_1(0) = 0$. The function $f_1(.)$ is defined in two terms of linear and nonlinear parts $f_1(x_1) = A_1 x_1 + \gamma(x_1)$. A_1 is the linearized term at the equilibrium point, and the nonlinear term $\gamma(x_1)$ contains higher-order terms such that $\lim_{\|x_1\| \to 0} \frac{\|\gamma(x_1)\|}{\|x_1\|} = 0$.

For system (12.16) the below sliding surface is defined:

$$\ell \overset{\text{def}}{=} \{x \in \mathbb{R}^n \ : \ s = c^T x = 0\}, c = \begin{bmatrix} c_1^T & 1 \end{bmatrix} \tag{12.17}$$

The SMC of nonlinear system (12.16) consists of two parts of the nominal controller and the switching part. Considering $\dot{s} = 0$ and $d = 0$ in nominal condition, u_{eq} is defined as follows:

$$\begin{aligned}\dot{s} &= c_1^T \dot{x}_1 + \dot{x}_2 = c_1^T f_1(x_1) + c_1^T B_1 x_2 + f_2(x_1.\ x_2) + B_2\, u + B_2 d \\ &= c^T f(x) + B_2 u_{eq} = 0\end{aligned}$$

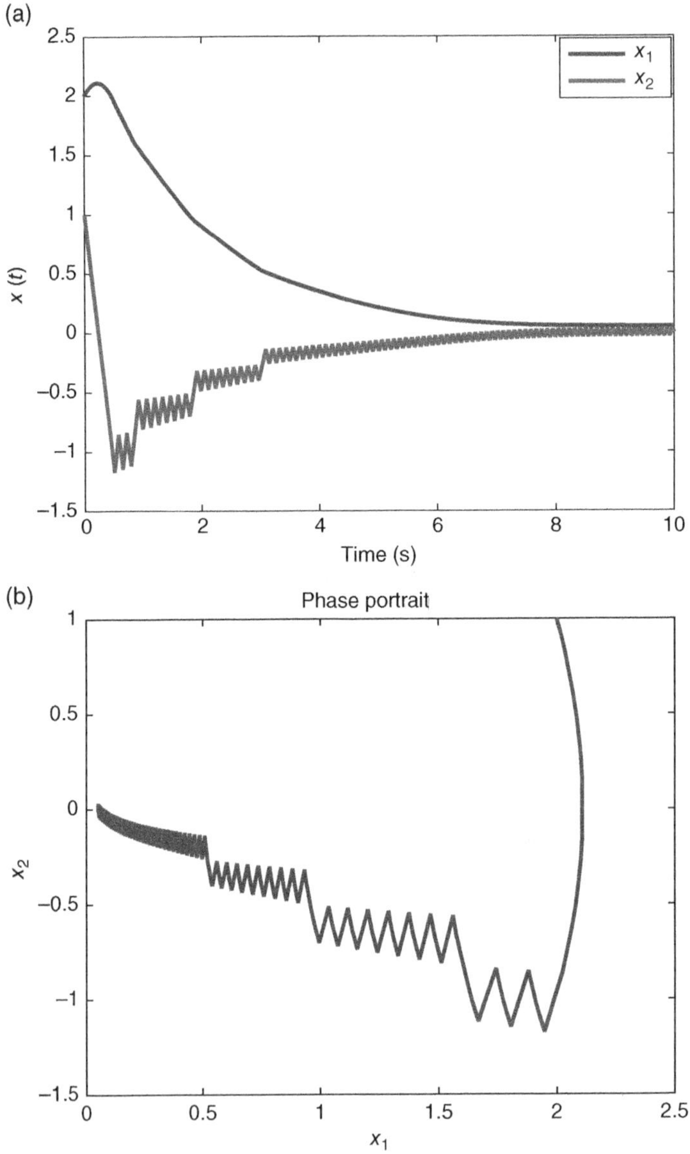

Figure 12.1 (a) States of LTI system is phase plane. (b) Phase portrait.

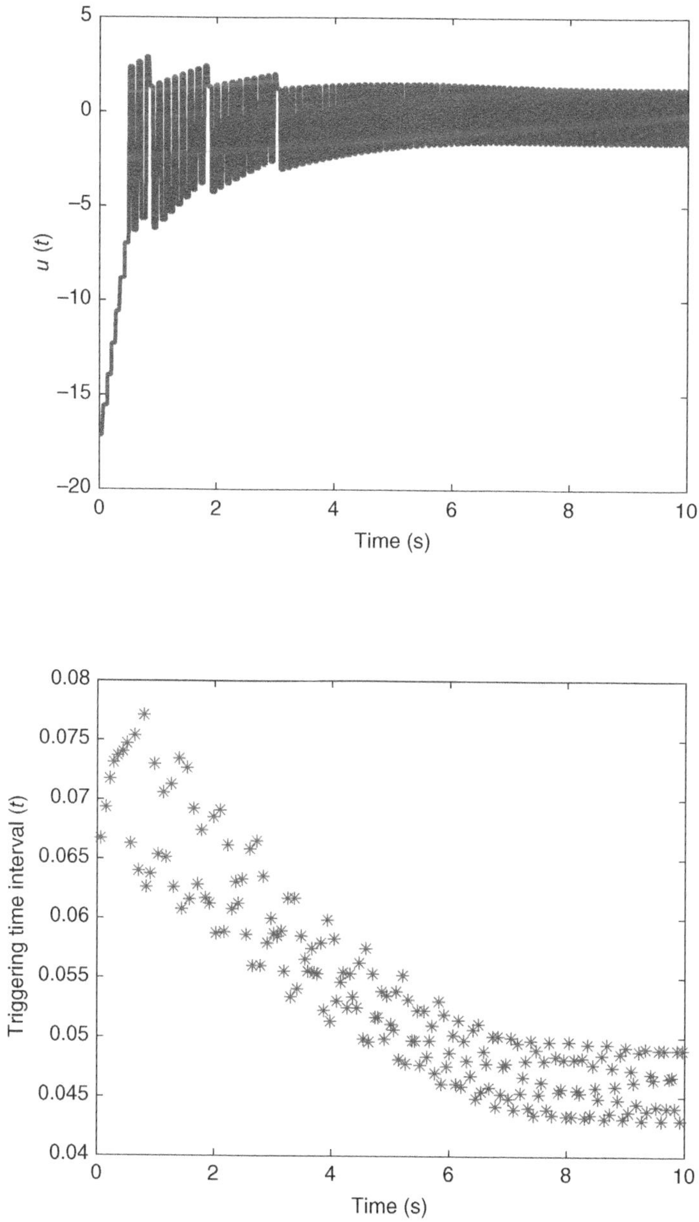

Figure 12.2 The controlling signal and inter-event times.

So it is concluded that

$$u_{eq} = -B_2^{-1}\left(c^T f(x)\right) \tag{12.18}$$

Adding the switching term yields:

$$u = -B_2^{-1}\left(c^T f(x) + K \operatorname{sign} s\right) \tag{12.19}$$

where according to Chapter 2, $K > |B_2| d_0$ confirms the finite-time reaching condition.

To design event-triggering SMC for system (12.17) the triggering time instants is $\{t_i\}_{i=0}^{\infty}$ where controller is considered to be constant in each time interval $T_i = t_{i+1} - t_i$. This is a kind of zero-order hold (ZOH).

So the event-triggering SMC is defined as below:

$$u = -B_2^{-1}\left(c^T f(x(t_i)) + K \operatorname{sign} s(t_i)\right) \tag{12.20}$$

Theorem 12.2
For system (12.16) controller (12.20) is a practical event-triggering SMC if for a positive constant $\alpha > 0$ which satisfies

$$L \, \|c\| \, \|e(t)\| < \alpha \tag{12.21}$$

there exists a K such that $K > |B_2| d_0 + \alpha$.

Proof: Considering the Lyapunov function

$$\dot{V}(s) = s\dot{s} \ = s\left(c^T f(x) + B_2 u + B_2 d\right)$$

and substituting controller (12.19) yields

$$\begin{aligned}
\dot{V}(s(t)) &= s(t)\left(c^T f(x(t)) - c^T f(x(t_i)) - K \operatorname{sign} s(t_i) + B_2 d(t)\right) \\
&\le -s(t) K \operatorname{sign} s(t_i) + |s(t)| \left\|c^T f(x(t)) - c^T f(x(t_i))\right\| + |s(t)| \, |B_2| \, d_0 \\
&\le -s(t) K \operatorname{sign} s(t_i) + |s(t)| \, \|c\| \, \|f(x(t)) - f(x(t_i))\| + |s(t)| \, |B_2| \, d_0 \\
&\le -s(t) K \operatorname{sign} s(t_i) + |s(t)| \, L \, \|c\| \, \|x(t) - x(t_i)\| + |s(t)| \, |B_2| d_0
\end{aligned}$$

Replacing the inequalities $L \, \|c\| \, \|e(t)\| < \alpha$, and $K > |B_2| d_0 + \alpha$ yields

$$\dot{V}(s(t)) \le -|s(t)| K + |s(t)| \alpha + |s(t)| \, |B_2| \, d_0 = -|s(t)|(K - \alpha - |B_2| \, d_0) < -\eta |s(t)|$$

□

The negativeness of $\dot{V}$ confirms the movement of systems trajectory toward sliding surface during time intervals $[t_i \ . \ t_i + 1)$. Although the above inequality cannot guarantee the uniform convergence of state trajectories, it leads to a bounded variation around sliding surface as follows:

$$\Omega = \left\{x \in \mathcal{D} : |s| = \left|c^T x\right| < \frac{\alpha}{L}\right\}$$

To investigate the practical event-triggering sliding mode control, the triggering scheme should satisfy the sufficient condition (12.21) [6].

So the triggering rule is considered as follows:

$$t_{i+1} = \inf\{t > t_i : L\|c\| \, \|e(t)\| \geq \sigma\alpha\} \tag{12.22}$$

For system stability, there must exist a positive lower bound between any two consecutive triggering instants to avoid Zeno execution of the sequence.

Lemma 12.1

Consider the system (12.16), controller (12.19), and the triggering rule (12.22). Then, the inter-execution time T_i is bounded as follows:

$$T_i \geq \frac{1}{L} \ln\left(1 + \sigma \frac{\alpha}{\|c\| \left(\rho_N(\|x(t_i)\|) + \beta_N\right)}\right); \tag{12.23}$$

where

$$\beta_N = \left\|B B_2^{-1}\right\| K + |B_2| \, d_0 \tag{12.24}$$

$$\rho_N(\|x(t_i)\|) = L\left(1 + \left\|B B_2^{-1} c^T\right\|\right) \|x(t_i)\| \tag{12.25}$$

Hint: For simplicity (12.16) is considered to be

$$\dot{x} = f(x) + Bu + Bd; B = \begin{bmatrix} 0^T & B_2 \end{bmatrix}^T$$

Proof: The time required for $\|e(t)\| = \|x(t_i) - x(t)\|$ to reach the value of $\sigma\alpha/\|c\|L$ is calculated as follows:

$$\begin{aligned}
\frac{d}{dt}\|e(t)\| &\leq \left\|\frac{d}{dt} e(t)\right\| = \left\|\frac{d}{dt} x(t)\right\| \\
&= \|f(x(t)) + Bu(t) + Bd(t)\| \\
&= \left\|f(x(t)) - BB_2^{-1} c^T f(x(t_i)) - BB_2^{-1} K \text{sign}(t_i) + Bd(t)\right\|
\end{aligned}$$

Using (12.24) and (12.25) yields:

$$\begin{aligned}
\frac{d}{dt}\|e(t)\| &\leq L\|x(t)\| + \left\|BB_2\left(c^T B\right)^{-1} c^T f(x(t_i))\right\| + \left\|BB_2^{-1}\right\|K + \|B_2 d(t)\| \\
&\leq L(\|x(t_i)\| + \|e(t)\|) + L\left\|BB_2\left(c^T B\right)^{-1} c^T\right\| \|x(t_i)\| + \left\|BB_2^{-1}\right\|K + |B_2| d_0 \\
&= L\|e(t)\| + \left(1 + \left\|BB_2^{-1} c^T\right\|\right) L\|x(t_i)\| + \kappa \\
&= L\|e(t)\| + \rho_N(\|x(t_i)\|) + \beta_N
\end{aligned}$$

Using Comparison lemma [7], and considering $\|e(t_i)\| = 0$, the above inequality leads:

$$\|e(t)\| \leq \frac{\rho_N(\|x(t_i)\|)\beta_N}{L}\left(e^{L(t-t_i)} - 1\right)$$

Considering (12.22) it is concluded that:

$\frac{\sigma\alpha}{L\|c\|} = \|e(t_{i+1})\| \leq \frac{\rho_N(\|x(t_i)\|)\beta_N}{L}(e^{LT_i} - 1)$ □

The inter-execution time should be greater than a given positive bound for T_i, because, in practice, the digital processors can only entertain the control task subject to some minimum time period corresponding to the processor bound width. So, Zeno execution is avoided in practical scenarios. In [5] a minimum value for α which guarantees that the inter-execution time T_i is larger than the processor bound width T_0 is proposed as follows:

$$\alpha_{min} = \frac{(\rho_N(\|x(t_i)\|) + \beta_N)\,\|c\|}{\sigma}(e^{LT_0} - 1) \tag{12.26}$$

According to above descriptions the event-triggering SMC for nonlinear system (12.16) is designed through the following steps:

Step1: Initialize σ, L, α_{min}, and T_0

step2: Select $\alpha \geq \alpha_{min}$ and calculate $K > |B_2| d_0 + \alpha$ and $\beta_N = \|B B_2^{-1}\| K + |B_2| d_0$

step3: Compute T_{0_est} from $\alpha_{min} = \frac{(\rho_N(\|x(t_i)\|) + \beta_N)\,\|c\|}{\sigma}(e^{LT_{0\_est}} - 1)$

step4: if $T_{0\_est} > T_0$

continue

else go to step2

step5: end

step6: $u(t) = -B_2^{-1}(c^T f(x(t_i)) + K\,\mathrm{signs}(t_i))$ $t \in [t_i, t_{i+1})$

$t_{i+1} = \inf\{t > t_i : L\|c\|\ \|e(t)\| \geq \sigma\alpha\}$

12.5 MATLAB Programming and Simulation Results

Consider a second-order nonlinear system

$$\dot{x}_1 = x_2$$

$$\dot{x}_2 = x_1 + x_2^2 + u + d; d = 0.5\cos(t)$$

$$\mathcal{D} = \{x \in \mathbb{R}^2 : \|x\|^2 = 9\},$$

$$L = 10, d_0 = 0.5, c^T = [0.5 \quad 1]; s = c^T x, x0 = -1.2, \sigma = 0.8$$

According to (12.20), the controller is considered to be

$$u = -B_2^{-1}\left(c^T f(x(t_i)) + K \text{ sign } s(t_i)\right)$$

Considering $c^T = [0.5 \quad 1]$ it is rewritten as:

$$u(t) = -x_1(t_i) - 0.5x_2(t_i) - x_2^2(t_i) - K \text{ sings}(t_i) \qquad t\epsilon[t_i, t_{i+1})$$

where according to (12.21) K is obtained from following inequalities:

$\alpha > 0$; $L \|c\| \|e(t)\| < \alpha$; $K > |B_2|d_0 + \alpha$.

The related MATLAB programming code is as follows:

```
%initialization
x0 = [-1; 2];
tf = 10;deltat = 0.0001;L = 10;
%system parameters
c = [0.5;1];
k = 2;
sig = 0.8;a = 0.5;%alpha

%initialize arrays
xindex = [];uindex = [];tiindex = [];IVindex = [];
%intereventindex
x = x0;xi = x0;
xindex = [xindex x0];
%sliding surface and controller initial values
    s = c'*x;
    u = -x(1)-(x(2))^2-k*sign(s);
    uindex = [uindex u];intereventindex = [];
%extracting event-triggered SMC
for i = 1: (tf/deltat)-1
    x = [deltat*x(2)+x(1);deltat*( x(1)+(x(2))^2+u+0.5*cos
(i*deltat))+x(2)];
    e = xi-x;
    lhs = L* norm(c)*norm(e);
    rhs = sig * a;
```

```
        if lhs >= rhs
        xi=x;
        s = c'*xi;
        ui = -xi(1)-(xi(2))^2-k*sign(s);
        u = ui;
        tiindex = [tiindex i*deltat];% the triggering time
        end
    uindex = [uindex u];
    xindex = [xindex x];
end
IVindex = [tiindex 0]-[0 tiindex]; %the interevent time is
ti+1-ti


%plot the results
t = 0:deltat:tf-deltat;
figure(1)
plot(t,xindex(1,:),t,xindex(2,:),'linewidth',2)
xlabel('t(s)');
ylabel('x(t)');
legend('x1','x2')
figure(2)
plot(xindex(1,:),xindex(2,:),'linewidth',2)
title('phase portraite');
xlabel('x1');
ylabel('x2');
figure(3)
plot(t,uindex(1,:),'linewidth',2)
xlabel('t(s)');
ylabel('u(t)');
figure(4)
plot(tiindex(1,:),IVindex(1,1:end-1),'*')
xlabel('t(s)');
ylabel('triggering time interval(t)');
```

Considering Figure 12.3, it is concluded that systems states effectively converge. And Figure 12.4 represents the controlling signal and inter-event time intervals.

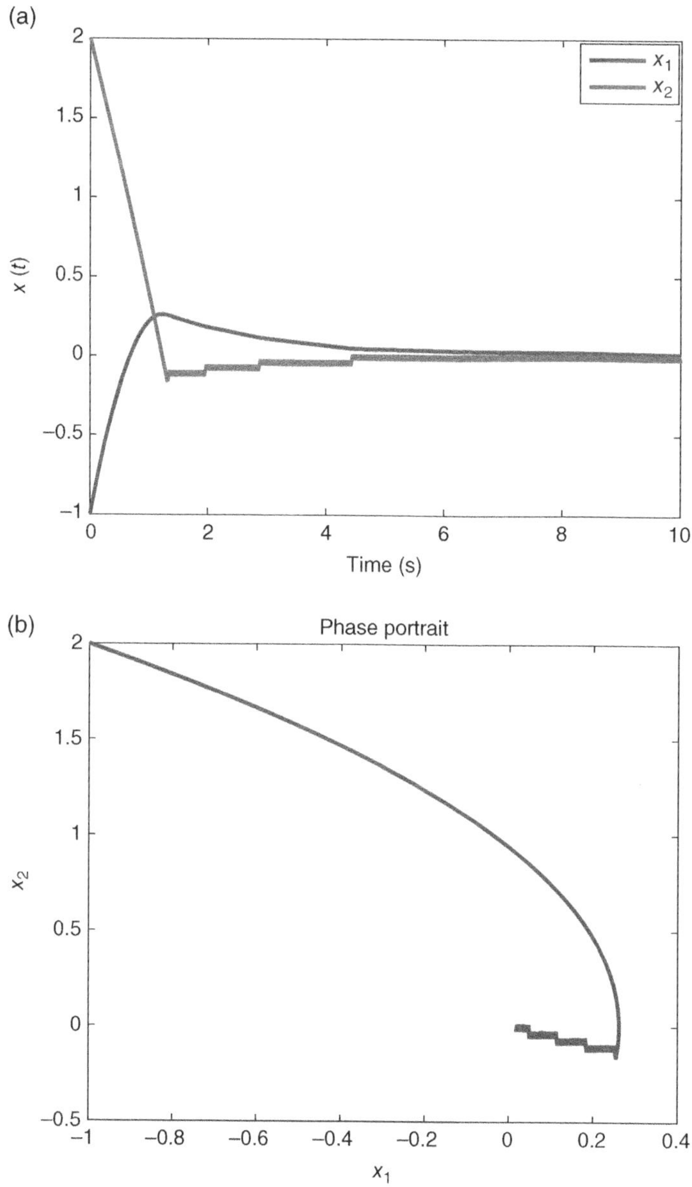

Figure 12.3 (a) States of nonlinear system is phase plane. (b) Phase portrait.

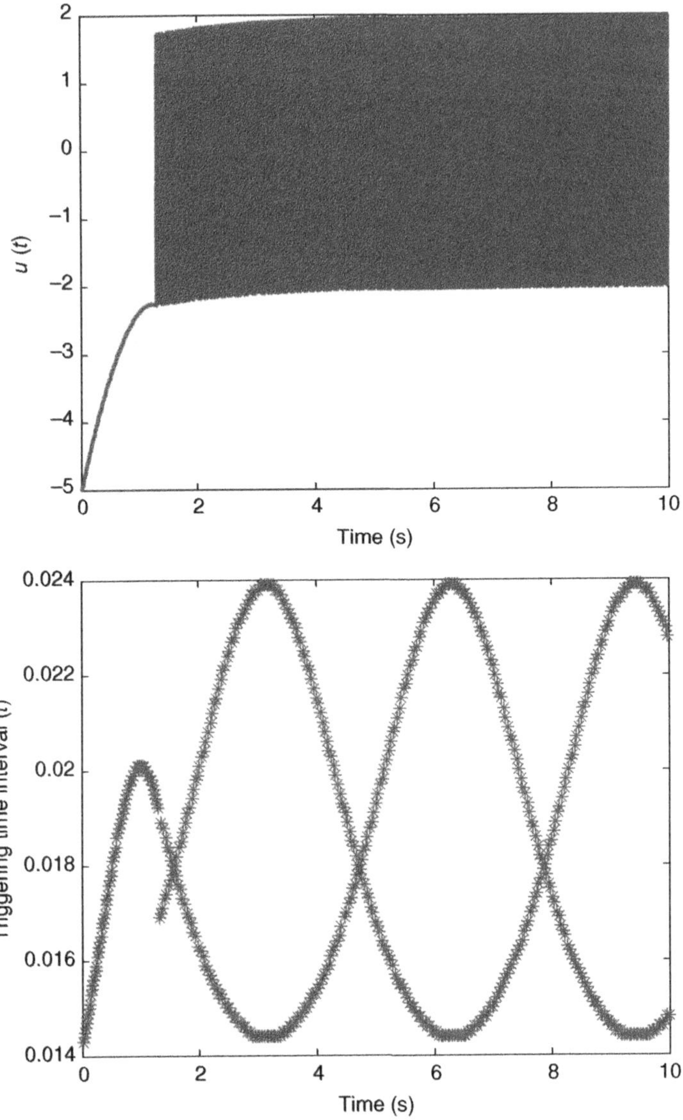

Figure 12.4 The controlling signal for nonlinear structure and inter-event times.

References

1 Li, W., Niu, Y., and Cao, Z. (2021). Event-triggered sliding mode control for multi-agent systems subject to channel fading. *International Journal of Systems Science* 53: 1–12.

2 Liu, J., Zhang, Y., Yu, Y., and Sun, C. (2020). Fixed-time leader–follower consensus of networked nonlinear systems via event/self-triggered control. *IEEE Transactions on Neural Networks* 31: 5029–5037.

3 Li, X., Peng, D., and Cao, J. (2020). Lyapunov stability for impulsive systems via event triggered impulsive control. *IEEE Transactions on Automatic Control* 65: 4908–4913.

4 Boundyopadhyay, B. and Behera, A.K. (2018). *Event-Triggered Sliding Mode Control A New Approach to Control System Design*. Springer.

5 Boundyopadhyay, B. and Behera, A.K. (2017). Robust sliding mode control: an event triggering approach. *IEEE Transactions on Circuits and Systems* 64: 146–150.

6 Liu, X.S., Shi, P., Shen, C., and Peng, Y. (2020). Event-triggered sliding mode control of nonlinear dynamic systems. *Automatica* 112: 108738.

7 Khalil, H. (2014). *Nonlinear Control*. Pearson.

Index

Methods of Developing Sliding Mode Controllers: Design and MATLAB Simulation, First Edition.
Reihaneh Kardehi Moghaddam and Mostafa Rabbani.

Published 2025 by John Wiley & Sons, Inc.

Books in the IEEE Press Series on Control Systems Theory and Applications

Series Editor: Maria Domenica Di Benedetto, University of l'Aquila, Italy

The series publishes monographs, edited volumes, and textbooks geared toward control scientists and engineers, as well as those working in various areas of applied mathematics such as optimization, game theory, and operations.

1. *Autonomous Road Vehicle Path Planning and Tracking Control*
   Levent Güvenç, Bilin Aksun-Güvenç, Sheng Zhu, and Sükrü Yaren Gelbal
2. *Embedded Control for Mobile Robotic Applications*
   Leena Vachhani, Pranjal Vyas, and Arunkumar G. K.
3. *Merging Optimization and Control in Power Systems: Physical and Cyber Restrictions in Distributed Frequency Control and Beyond*
   Feng Liu, Zhaojian Wang, Changhong Zhao, and Peng Yang
4. *Dynamic System Modelling and Analysis with MATLAB and Python: For Control Engineers*
   Jongrae Kim
5. *Model-Based Reinforcement Learning: From Data to Continuous Actions with a Python-Based Toolbox*
   Milad Farsi and Jun Liu
6. *Disturbance Observer for Advanced Motion Control with MATLAB/Simulink*
   Akira Shimada
7. *Control over Communication Networks: Modeling, Analysis, and Design of Networked Control Systems and Multi-Agent Systems over Imperfect Communication Channels*
   Jianying Zheng, Liang Xu, Qinglei Hu, and Lihua Xie
8. *Advanced Control of Power Converters: Techniques and MATLAB/Simulink Implementation*
   Hasan Komurcugil, Sertac Bayhan, Ramon Guzman, Mariusz Malinowski, and Haitham Abu-Rub
9. *Sensorless Control of Permanent Magnet Synchronous Machine Drives*
   Zi Qiang Zhu and Xi Meng Wu
10. *The Impact of Automatic Control Research on Industrial Innovation: Enabling a Sustainable Future*
    Silvia Mastellone (Editor) and Alex van Delft (Editor)
11. *Interval Analysis: Application in the Optimal Control Problems*
    Navid Razmjooy
12. *Dynamic Modeling and Neural Network-Based Intelligent Control of Flexible Systems*
    Hejia Gao, Wei He, and Changyin Sun
13. *Methods of Developing Sliding Mode Controllers: Design and MATLAB Simulation*
    Reihaneh Kardehi Moghaddam and Mostafa Rabbani

Printed and bound by CPI Group (UK) Ltd, Croydon, CR0 4YY
17/09/2026
14964234-0001